HISTORY, PHILOSOPHY AND SOCIOLOGY OF SCIENCE

Classics, Staples and Precursors

HISTORY, PHILOSOPHY AND SOCIOLOGY OF SCIENCE

Classics, Staples and Precursors

Selected By

YEHUDA ELKANA
ROBERT K. MERTON
ARNOLD THACKRAY
HARRIET ZUCKERMAN

PHILOSOPHY AS SCIENTIA SCIENTIARUM

AND

A HISTORY OF CLASSIFICATIONS OF THE SCIENCES

BY

ROBERT FLINT

ARNO PRESS
A New York Times Company
New York – 1975

Reprint Edition 1975 by Arno Press Inc.

Reprinted from a copy in
The University of Illinois Library

HISTORY, PHILOSOPHY AND SOCIOLOGY OF SCIENCE:
Classics, Staples and Precursors
ISBN for complete set: 0-405-06575-2
See last pages of this volume for titles.

Manufactured in the United States of America

Library of Congress Cataloging in Publication Data

Flint, Robert, 1838-1910.
Philosophy as scientia scientiarium.

(History, philosophy, and sociology of science)
Reprint of the 1904 ed. published by W. Blackwood, London.
1. Classification of sciences. 2. Philosophy.
I. Title. II. Title: A history of classifications of the sciences. III. Series.
BD241.F6 1975 111 74-26261
ISBN 0-405-06589-2

PHILOSOPHY AS SCIENTIA SCIENTIARUM

AND

A HISTORY OF CLASSIFICATIONS OF THE SCIENCES

PHILOSOPHY AS SCIENTIA SCIENTIARUM

AND

A HISTORY OF CLASSIFICATIONS OF THE SCIENCES

BY

ROBERT FLINT

D.D., LL.D., F.R.S.E.

CORRESPONDING MEMBER OF THE INSTITUTE OF FRANCE;
HONORARY MEMBER OF THE ROYAL SOCIETY OF PALERMO; AND
PROFESSOR (EMERITUS), EDINBURGH UNIVERSITY

WILLIAM BLACKWOOD AND SONS
EDINBURGH AND LONDON
MCMIV

PREFATORY NOTE.

VERY little requires to be referred to here. The Table of Contents should render unnecessary any Index of Names. The author's connection with the subject of his book has been a lengthened one. When a mere youth in Glasgow University he joined a number of young men, among whom were representative Canadians, Englishmen, Welshmen, and others, in forming a Literary and Philosophical Society. As a member and vice-president his contributions to it were two essays, one on "Cartesianism" and the other on "The Relations of the Sciences." The former cost him a study of two hundred old books in Latin and French, but it soon got lost and never returned to him. The latter he still possesses, and deems on the whole fairly accurate so far as it goes. His dealing with such a subject at all he attributes to the inspiration of the greatest of his teachers, the Professor William Thomson of the time, the Lord Kelvin of to-day and of all time. My study on the "Relations of the Sciences" did not deal at all with the history of the subject, but kept entirely to what was implied in the title. During some years I was entirely engrossed with pastoral duties. In 1864-65, my first session as Professor of Moral Philosophy and Political Economy at St Andrews, I gave some lectures on the connection of those two sciences to other sciences. In 1867 I had begun to think of constructing an elaborate work on the Relations of the Sciences to one another, to Philosophy, Religion, and Morality, and such a work was advertised for a considerable number of years. The delay and revocation must have been hard on the publishers, but I suppose publishers get accustomed to such things. For myself I deem it fortunate and even providential to have had to change my intended course and follow others where more urgent demands were made and more obvious interests were at stake. A considerable portion of the *History of Classifications of the Sciences* appeared in America long prior to any portion of it in Britain. The portion referred to will be found in the July number of the *Presbyterian Review* for 1885. Dr Briggs and Dr Patton were the chief editors of the Review. Dr Calderwood, Dr Blaikie, Dr Croskery, and I were associate editors for Great Britain. Never can I forget the kindness and worthiness of them all. Alas! few of them now remain here below. May those of us who are still here walk worthy of those who have gone before.

R. FLINT.

1 Mountjoy Terrace, Musselburgh, N.B.,
24*th September* 1904.

CONTENTS.

PHILOSOPHY AS SCIENTIA SCIENTIARUM

PHILOSOPHY AS SCIENTIA SCIENTIARUM.

I.

THE sciences are parts of a great whole, the members of a magnificent system. Each of them has manifold relations to every other. But the great whole, the magnificent system, to which they belong is itself an object of knowledge. Unless the intellectual universe be no real universe, but essentially a chaos, science must be general as well as special; or, in other words, there must be a science of the sciences—a science which determines the principles and conditions, the limits and relations, of the sciences. This science is philosophy; and what the author has to say in the present chapter is meant to be a plea for philosophy as the legitimate but often disavowed and insulted queen of the sciences. "Time was," says Kant, "when metaphysics was the queen of all the sciences. But now it is the fashion to heap contempt and scorn upon her, and the matron mourns, forlorn and forsaken, like Hecuba." The sciences, however, cannot do without a queen. There may be a republic of letters, but the sciences cannot constitute a republic; they must be so connected as to form a unity; and the science which refers them to unity and shows that knowledge as a whole is a cosmos is the supreme science, the queen of the sciences. The want of practical recognition of this truth is one main cause of the intellectual anarchy of our times.

Philosophy as scientia scientiarum may have more functions than one, but it has at least one. It has to show how science is related to science, where one science is in contact with another, in what way each fits into each, so that all may compose the symmetrical and glorious edifice of human knowledge, which has been built up by the labours of all past generations, and which all future generations must contribute to perfect and adorn. With whatever province of science a thoughtful man occupies himself, he soon becomes aware that it has intimate and manifold connections with other provinces, and if he try to trace these connections out, he will ere long perceive that the sciences are not isolated things, but so bound together as to constitute a unity which is a reflection of the unity of nature and of the unity of that Supreme

Reason which pervades all nature and originates all intelligence. Philosophy aims to raise the mind gradually and legitimately to a point from which this unity may be visible, while the distinctions of the special sciences are not only not effaced, but lie clearly and truthfully before it. If I seek to vindicate and magnify this aim it is not because I suppose its reasonableness is likely to be directly and explicitly denied, but because its importance can scarcely in the present day be too often or strongly insisted on. There is many a truth which is not contested, which receives a ready acquiescence of a sort, and yet which is very far from being apprehended or generally acted on, because the evidence for it is not so definitely and adequately before the mind as to counteract influences which tend to obscure it and make it practically neglected. And that aspiration after insight into the system of science as a whole should not be lost in the study of details is pre-eminently such a truth.

Now, the first consideration which here suggests itself is that philosophy, viewed as scientia scientiarum, is simply science which has attained to a knowledge of the unity, self-consistency, and harmony of the teachings of the separate sciences. Philosophy seeks to do for the sciences just what each science does for the doctrines it comprehends.

In the latter case separate truths are brought into unity, and in the former separate sciences. The one unity constitutes a science, the other a science of the sciences, and shows that absolutely there is but one science, although it has various departments, whereby the incommensurableness of nature is brought down to our capacities. The second and higher unity is as natural, as legitimate, as important as the first and lower unities. It would little avail, indeed, that these existed—that there was unity enough in things to permit of the formation of special sciences—if there were no still more comprehensive unity, if the point of view of each science was in itself final, if each science was utterly isolated from all others. If such were the case there would be in science something essentially disappointing to the human mind, for it would be of its very nature calculated not to satisfy but to thwart that love of unity which is the source and life of all scientific research. If such were the case truth would not form a fair and harmonious body, but it would resemble the mangled and scattered limbs of Osiris, while the human mind in its pursuit would be engaged in a task more mournful than that of Isis, because hopeless. It is not so, however, but

> "The One through all in cycles goes,
> And all to One returning flows."

Science is not sectioned into entirely unconnected sciences. In all the sciences there is a certain common nature, and among them there are many ties of affinity and points of contact. There are precedence and subordination, order and harmony, among them; so that, many and diverse as they are, they form a whole, a system in which each of them has its appropriate place, and, so far from being sacrificed to any other, has a new dignity imparted to it by being referred to the final unity of reason, the common centre of knowledge.

Secondly, philosophy, as a comprehensive survey of the sciences and a deeply grounded knowledge of their principal relations to one another, is a condition indispensable to a correct conception of the special province of any science. The boundaries of most sciences are very ill-traced, their definitions most irreconcilable. The first question which the student of any science naturally asks, What is it? What is it about? is one to which he can often get no satisfactory answer—one on which he finds that all the doctors disagree. Take logic. One logician will tell you its proper object is thought as thought; another, that it is the forms as contradistinguished from the contents or matter of thought; another, that it is only the necessary as distinct from the contingent forms of thought;

another, that it is only a kind of thought, mediate or discursive thought; another, that it is only a kind of mediate or discursive thought—inference; and still another, that it is not thought as thought, nor any elements or kinds of thought, but qualities of thought—truth and error so far as involved in the application of thought. And, it must be remarked, this opposition is in no way one between old and new views, between transcended and effete conceptions and those which actually prevail, but one which exists between the most deliberately formed convictions of the most eminent modern logicians. Certainly it is a somewhat perplexing puzzle to lie at the very entrance of a science. The ingenuous youth who makes his first acquaintance with logic by getting that nut thrust into his mouth is not likely, if his teeth be sharp enough to crack it, to find any subsequent problem too hard for him. It is not much otherwise with psychology, with rhetoric, with ethics, with politics, with political economy. And as to metaphysics, it fares far worse; the discordance and embroilment there baffle description, for, as Professor Ferrier so happily said, "All the captains are sailing on different tacks, under different orders, and under different winds; and each is railing at the others because they will not keep the same course with himself. One man is playing at chess, his adversary is playing against him at billiards; and whenever a victory is achieved or a defeat sustained, it is always such a victory as a billiard-player might be supposed to gain over a chess-player, or such a defeat as a billiard-player might be supposed to sustain at the hands of a chess-player."

Now, how is such a state of things to be remedied? How are we to decide between the disputants? How make a choice for ourselves between conflicting definitions? It is obvious neither tradition nor authority can here help us, for not only are they in themselves discordant and undecided, but they have no right to overrule reason, which ought to submit to evidence alone, and is unworthy of itself when it listens to any other voice than that of truth. Nor will it suffice to found our definitions on the etymology and inherent significance of names. That may wholly mislead. Words often come to signify what is altogether different from their intrinsic meaning, sometimes what is the reverse of it. A manufacture, for instance, is not what is made by the hand, but what is made by machinery with little or no aid from the hand. Words may be stretched or contracted, where needful, to conform to realities, but realities are not to be twisted in any way to conform to words; and it is not with words but realities that science has to deal. It may be said, a science cannot be defined until after the study of its appropriate facts, and when the study is sufficiently advanced the definition comes of itself. And that is so far true. Although first in the order of exposition, the definition of a science is late in the order of discovery and presupposes a certain acquaintance with an appropriate order of facts, expressing, as it does, some essential characteristic which they all possess. But the question is, the difficulty is, to determine what is the appropriate order of facts, why the one chosen and not another, why an order of a given extent instead of one larger or smaller. All the views of logic, for instance, to which I have referred assign to it a natural order of facts, a sphere of real knowledge worth acquiring, a sphere with distinct enough boundaries; and yet the natural orders are not coincident, the boundaries are altogether different, some going all round those of others, and others intersecting one another in the most perplexing ways.

Now, in such a case, it is obvious there is but one mode of deciding who is right and who is wrong, who has selected the proper group of facts and who groups larger or smaller, who has traced the boundaries of his science well and who ill. It is by examining whose views give to their science a place that fits in rightly into the scheme of science. The question is one of adjustment. The logician simply as logician cannot define logic, for that is an affair of the settlement of boundaries between the sort of knowledge he cultivates and contiguous divisions of knowledge, such as metaphysics, psychology, and rhetoric; one, accordingly, which can only be decided by a higher and more general sort of considerations than belongs to any special science—by considerations as to the relations of the sciences. And this holds universally. It is as impossible to fix the position of a science without reference to neighbouring sciences, and even to the general system of the sciences, as to fix the position of a nation without reference to surrounding nations, and even to the general geography of the earth. In this respect a general scheme of science is exactly like a general map or like a terrestrial globe; and like such map or globe it supplies a want which can no otherwise be provided for. An atlas with a separate map of every state in the world cannot dispense with, cannot supply the place of, a map which will show them in relation; nay, the more complete an atlas is in special maps the more need is there of a general one, because the more certainly and the more deeply will the student without such assistance be lost in details. And so with respect to science. The more it becomes divided and sub-divided, the more urgent, the more imperative becomes a knowledge of its greater general outlines in order that each man may recognise how the department he is specially conversant with is related to others. The greater the multiplication of sciences the more chaotic must be the effect they produce unless the mind can locate them aright, can refer them to their place in a system, and see how they stand to one another and the whole.

What has now been said leads to a third consideration in favour of philosophy as viewed from our present standpoint. By a true co-ordination of the sciences and a comprehensive insight into their natures, it must help us to see how and when they can assist each other. There are problems which require a combination of sciences for their solution; there are certain combinations of the sciences possible, while others are absurd; and it is only through a clear apprehension of the respective natures and relations of any two or more sciences that we can perceive if one can be made to operate with another to the attainment of a given end. Some of the most important advances which have occurred in the history of science have been due to the associated action of two or more sciences. A signal instance is Descartes' application of the algebraic analysis to define the nature and investigate the properties of curve lines. It was only by the clearest conception of the relations of the two sciences, algebra and geometry, that he could have brought the symbols and calculations of the one to bear on the problems of the other, and thus start a new epoch in mathematical science. A more modern instance of the same kind is the union of chemistry and optics in spectral analysis, by which the most singularly interesting results as to the physical constitution of the heavenly bodies have been attained. It will be in the future as it has been in the past. Some of the most difficult and important of the problems which are at present attracting the curiosity and trying the ingenuity of men can only, it is apparent, find their solution from a happy combination of chemistry and physiology; others still more vital only from the combination of physiology and psychology; and not a few are so complex that it is vain to hope that they will be mastered otherwise than by the conjoint and concentrated efforts of many sciences. It is most erroneous to suppose, as some persons do, that the true way to advance any study is to devote the whole mind exclusively to it so as to have no thought or interest beyond it.

The sciences advance by solving problems which are often presented to them from without, and by accepting hints and helps from all sides. Mathematics itself, although it has in the character of its fundamental conceptions an enormous advantage over all other knowledge as abstract science, has found its chief stimulus in the requirements of the natural philosopher, in the problems of astronomy, mechanics, optics, heat, and electricity. "The combinations arising out of external phenomena," said Principal J. D. Forbes of St Andrews, "are more suggestive of the possible relations of number and quantity than is the most unlimited stretch of fancy and imagination." And if even mathematics, which is based on such singularly simple, precise, definable, workable conceptions as number and quantity, thus needs light from without, and only prospers because readily responsive to external suggestions, what can be expected from, say, logic, psychology, or ethics, which have vastly vaguer conceptions to start from, attempting to proceed entirely from within, and ignoring the combinations of human nature which are presented to us in history, in literature, and in language; what but that which we not unfrequently see—men working their way laboriously and painfully into a world of mere formulæ, of words and nothing but words, although doubtless big and brave words—a region of absolute emptiness, into which we may as well not follow them, however much we may admire the strength of constitution which enables such privileged natures to sustain life in a vacuum?

Whatever may be fancied to the contrary, the

truth is that the researches and studies of the mere specialist are never very productive. Special investigations only enrich science to any considerable extent when they are directed and guided by enlarged views; they are only truly successful when not exclusively special; when, on the contrary, the part or section of existence examined is looked at by a reason illumined by a worthy and ample idea of science; a reason which sees the part in the light of the whole and the whole as related to the part. I do not deny that now and then, by a lucky chance, a mere specialist may come across something valuable; that an entomologist who has no interest in anything but beetles may detect something in the eye or on the wing of some of these creatures which wiser men than himself can turn to good account; or that the most unintelligent local antiquarian may not find in some old document or mound or ruin a fact which decides the fate of a brilliant historical hypothesis: but I do affirm that discoveries thus made are extremely rare. Have not the most minute researches of recent botanists, zoologists, physiologists, &c., had reference to the vast generalisations and bold conjectures of a Spencer and a Darwin? What special historical researches have ended in the adequate solution of a complicated and difficult problem, except those conducted by men whose insight into the general providential plan of history, or at least of a large portion of history, was clearer and more profound than that of other men? I know of none. Now, what does all that amount to, but just that a study, a science, is progressive and flourishing only in so far as it is impelled and guided, penetrated and pervaded, by the spirit of philosophy; that all scientific discoveries whatever lie in the path along which philosophy leads science—along which science tends towards philosophy?

Philosophy, understood as has been explained, is, I remark fourthly, fitted and needed to counteract the evil intellectual and moral influences of specialism. We are all narrow by nature, and we require to have our narrowness guarded against and corrected, not confirmed and intensified. Different minds have different natural aptitudes. These different aptitudes find their appropriate spheres of exercise in special studies and special departments of practical life. A man with a genius for languages may have no turn for mathematics. The born poet may be the reverse of specially qualified for success either in science or business. The shrewdness and decision of mind which go so far to ensure success in the commercial world are useful gifts anywhere, but will certainly count for less in the world of learning than of traffic.

intellectual life, although we cannot attain to all truth, we ought carefully to cherish the love of all truth. But this, I need hardly say, is very difficult to do in the present state of society, when the division of scientific as well as of industrial functions is extreme.

A great and thoughtful poet, struck with the obvious and terrible dangers which, in consequence, threaten the spiritual life, has said :

> ". . . . Go demand
> Of mighty nature if 'twas ever meant
> That we should pry far off and be unraised,
> That we should pore, and dwindle as we pore,
> Viewing all objects unremittingly
> In disconnection dead and spiritless ;
> And still dividing, and dividing still,
> Break down all grandeur, still unsatisfied
> With the perverse attempt, while littleness
> May yet become more little : waging thus
> An impious warfare 'gainst the very life
> Of our own souls."

Now truth and error are mingled there and must be separated. It *was* meant by mighty nature that we should go on, as we have been doing, " still dividing, and dividing still "; it *was* meant that we should break down all grandeur into its constituents ; that the life which we cannot create we should yet in order to understand dissolve into its elements and view them unremittingly, " dead and spiritless" although they be; that we should be unsatisfied "while littleness may yet become more little," while division has not reached its utmost limits, while analysis has anything more to do. Division, analysis, is a necessary and inevitable condition of progress both in life and science. Every stage of progress must be consequent on a stage of division, spontaneous or reflective, industrial or scientific. We can well forgive a poet being slow to believe in the existence of such a law; but the law exists, and it will not avail us to ignore it, still less to resist it. This law, however, like every other, requires to be watched and its incidental evils guarded against. It is not more true that it is one of the conditions on which the progress of science and the advancement of society depend, than that if left to itself, if not balanced and counteracted by other agencies, it will arrest science and destroy society. But nature has provided forces with which it has only to be rightly adjusted in order that its action may be purely beneficial. If in one respect the subdivision of industrial labour has a narrowing and anti-social influence, it has in the other respect, that it condenses population within narrow circuits, associates intelligences and forces, and multiplies the objects of common interest, as well as the occasions for sympathy and the facilities for education, an influence altogether contrary, which has only to be made the most of and secured to the side of truth and goodness in order that all the evils incident to the specialisation of functions in modern industry may be scarcely recognisable when laid by the side of its benefits. In a general doctrine of science, the expression of that pure love of truth in its entirety which is identical with the spirit of philosophy, there is no less obviously a natural remedy for the evils incident to the specialisation of the sciences. Such a doctrine would enable the specialist to transcend the bounds of his own department, to realise his relation to science as a whole, and his own relation to all his fellow-labourers in science. Limited as his own particular study might be, it would no longer be a something " dead and disconnected," but united to the ultimate principles which are the root of all science, and through that union filled with the life which the root alone supplies.

This leads me to remark that philosophy, thus viewed, would afford the most important guidance in education. It must be, indeed, the very basis of rational education in science. It must be what best determines the course to be pursued. We cannot commence the study of science at any point nor prosecute it in any order we please. Nature has determined both where we ought to begin and what path we ought to follow. It is very far from a matter of indifference which of the mathematical sciences we commence with. If we plunge into natural philosophy without any mathematics to buoy us up we are likely soon to repent of our foolhardiness, and are certain not to swim very far. We shall make a similar mistake if we enter on moral philosophy without having made ourselves acquainted with the leading truths of psychology. Now, a philosophy of science worthy of what it should be would inform us at once what science was the natural antecedent of any other science, the condition of its intelligibility. It would, in fixing the order of the sciences, fix likewise the order of their rational study. It would thus lay what is the very corner-stone of the science of education — that without which no such thing as a science of education can exist. And it would confer on education another advantage only inferior to that. It would show what science was most fitted to correct the mental vices generated by any other science, as well as what science was needed to render it intelligible. No one science does more than cultivate the mind in a partial and one-sided manner; and if we would have fully developed, well-balanced minds, we must not only not confine ourselves exclusively to one, but counteract that which is exclusive and hurtful in our special pursuit by the kind of knowledge most unlike it in character and tendencies; that which it requires the most directly opposite procedure of mind to appropriate; that which exercises with most intensity the faculties which the other leaves most dormant.

Those who cultivate a science which is entirely inductive, which is only in process of formation, still unsettled in its foundations, still vague and dubious in the majority of its conclusions, while they can need no mathematics merely to render it intelligible, are precisely those who will need most the peculiar discipline of mathematics; and without it their power of deduction will remain unexercised; without it the very notion of what complete proof is will never find a place in their minds. On the same principle, the study of physics and psychology should be conjoined in one culture. The one is required to balance the other. All physicists should seek a general acquaintance with psychology, and all psychologists a general acquaintance with physics. This would remove the unbecoming antagonism which has so long and widely prevailed between those two classes of students — an antagonism which has its origin in ignorance, and is a signal proof of the narrowness of intellectual conception and illiberality of feeling which are produced by specialism when left to operate without check or counterpoise. This, then, is also to be said on behalf of a science of the sciences, that it would at once and authoritatively tell where the knowledge requisite to condition or the knowledge requisite to supplement or balance any other knowledge was to be found.

I now come to a consideration at least as weighty as any of those which have already been mentioned —namely, the interest and importance of the truths with which a science of the sciences must be conversant. The truths which lie *between* the sciences are as real and have equal claims to attention as the truths *within* the sciences. If the relations between facts are as important as the facts themselves,—and every science acknowledges and proceeds on this assumption,—how should the relations between the sciences not be of extreme interest and value? When these relations are known, all the facts any given special sciences deal with, and all the laws which have been derived from these facts, have a new light shed on them by being connected, contrasted, and compared from an elevation which permits of a truthful survey. That the relations of the sciences to one another are in themselves most worthy of examination, any one may convince himself by considering for a moment what they are, what great problems they present, what grave interests they involve. How are the mathematical sciences related to one another and to physics? Do they originate in experience, or are they offshoots of a transcendental or metaphysical condition? Are there any limits in nature to their application, and, if so, what are those limits? These are questions which mathematics suggests, although it does not solve,—hard and abstruse but real and not fanciful, weighty and not trivial questions,—and on which not philosophers only, but men whose distinctions have been gained chiefly in mathematics, such as Courtot, Sir Wm. R. Hamilton of Dublin, Boole, De Morgan, Bartholmai, Duhamel, have written either books or elaborate essays. How are the physical sciences related? Which are simple and fundamental, which complex and applicate? What must each take from others, and what may each be made to contribute to others? These, again, are questions which all physicists, not dwarfed by exclusive specialism of pursuit into incapacity of large views of any kind, are keenly alive to; for they see that on clear and correct views regarding them the future progress of physical science is greatly dependent, and a right settlement of the practical problem, What is a wise and well-conducted education in physical science? entirely dependent. What is the relation of the physical to the mental sciences, or even merely, What is the relation of physiology to psychology? No man can be so intellectually blind as to fail to perceive what a most momentous question this is. Every thinking man must answer it in some form or way; yet if you answer it in one way you must be a materialist, if in another a pure idealist, and it is to be hoped that it can be answered also in a third way which will make you neither—which will not compel you, as a rational being, to deny the existence either of matter or spirit, either of your bodies or souls.

Then, as to the mental sciences, psychology, ethics, æsthetics, politics, paideutics, philology, philosophy of history, &c., nothing is more certain than that a very large proportion of the evils which infest them, and which have given such abundant occasion to their adversaries to misrepresent and depreciate them, are due precisely to the want of definite and correct views in their cultivators as to their boundaries and relations; so that inquiries proper to one have been inextricably mixed up with inquiries proper only to another, and not unfrequently even this has been aggravated and confusion itself still further confounded by the introduction of the still more extraneous elements of physics, and metaphysics, and religion.

There is not less involved in the question, How is metaphysics related to physical and mental science? There are those who suppress metaphysics entirely, who regard it as only an erroneous phase of thought, gradually drawing near to the death which is its doom,—who maintain that there is no science save realistic or positive science. There are others who, instead of thus absorbing metaphysics in positive science, have sought to absorb all positive science in metaphysics, pretended to "re-think the great thought of creation," and hesitated not to deny the law of gravitation, to blame the very stars, to pronounce the most ancient heavens wrong, when these things did not appear to conform to their deductions. And between these two extremes, the Comtist and the Hegelian, there are innumerable other erroneous positions, into any of which it is easy to fall; while to get sure footing on the one right spot no man can, unless by working out for himself a correct and adequate apprehension of the relation of metaphysics to experience.

Quite as important as the question just referred to is this other question, How are piety and knowledge, religion and philosophy, theology and the physical and mental sciences, to be shown in their true relationship? Even in this age of many wants there are few, if any, more to be desired than a right answer to that question. The false and mischievous attitudes so often assumed by scientific men towards religion and by religious men towards science may unquestionably be largely traced to such erroneous conceptions of the relationship between religion and science as can only be dispelled by a thorough and unprejudiced philosophical investigation. And it can hardly be in the power of man to render better service to either religion or science than to exhibit them in their true natures and relationships, seeing that both of them, and society as well, are so grievously suffering from the want of clear and just views on the subject.

II.

In seeking to attain self-consistency and completeness philosophy must strive to solve four very comprehensive and complex problems.

In the first place, it has a duty towards the special sciences.

It is bound to form a right estimate of them and to take up a right attitude towards them. It is science, yet not merely a special science, but the science which has the processes and results of all the special sciences for its data—the general or universal science which has so risen above the special and particular in science as to be able to contemplate the sciences as parts of a system which reflects and elucidates a world of which the variety is not more wonderful than the unity. Philosophy should neither attempt to do the work nor to dispense with the aid of any special science, but must seek so to understand the methods, to appreciate the findings, and to trace the relationships of all the special sciences as to be able to combine them into a harmonious *cosmos* or well-proportioned *corpus.* When engaged in this task it may appropriately and usefully, perhaps, be called *positive philosophy,* and nearly corresponds to what has been so designated by Comte.

Comte's view of philosophy, however, as merely a generalisation of the results of the sciences, would have been an inadequate one even if he had duly recognised the existence and claims of the psychological and theological sciences. It is necessary to hold to the truth which is in Kant's view, and to the truth which is in Ferrier's or Hegel's view, of the nature of philosophy, quite as firmly as to the truth which is in Comte's view. Given a complete knowledge of the relations of the sciences—given, consequently, a correct picture on the mind's eye of the whole intelligible world drawn from the highest and best established results of all the sciences—and the work of reason, which is the comprehension of itself and of its objects so far as knowable, is still far from accomplished; yea, its highest and perhaps hardest labours have not yet begun. Scientific thought is not necessarily self-criticising thought; on the contrary, mere scientific thought, however rigid and methodical, is essentially dogmatic thought in that it rests on untested and uncriticised assumptions. It is reasoned, yet unreflective. It builds up what is currently admitted to be knowledge, but it does not inquire what so-called knowledge is or is essentially worth. The philosophy which wholly depends on such assumed thought or knowledge has all their essential defects. It is merely an advance on special science, as special science itself is on ordinary knowledge, and ordinary knowledge on crude sensation. Along the whole line the mind never changes its attitude towards its objects. At the end its nature is just what it was at the beginning. Throughout what it brings with it is borrowed ordinary knowledge or positive science. The scientist often fancies that he is a man who takes nothing on trust when in reality he takes everything on trust, because he accepts without question or reservation thought itself as naturally truthful and its laws as valid. Whatever superficial scientists may suppose to the contrary, the fact is that the entire procedure of science and of philosophy, in so far as it is simply a generalisation of science, is assumptive and dogmatic. Although often contrasted and opposed to faith it really rests on faith, and in the view of a serious and consistent scepticism must rest on blind faith.

Thought may assume, however, and is bound to assume, a very different attitude towards itself and towards its objects. It may pass and ought to pass *from a believing to an inquiring, from a dogmatic to a critical stage.* It may turn its attention and force from a study of the relations of the known to an examination of the conditions and guarantees of knowledge.

In the second place, then, philosophy is bound to institute an investigation into the nature of knowledge itself.

All the special sciences aim merely at the extension and acquisition of knowledge. They assume that there are things and truths to be known, but make no attempt to verify the assumption or even to understand what it implies. What are things apart from knowledge and in relation to knowledge? Are things just what they appear to be, or not at all what they appear to be, or partly what they appear to be and partly not? May all things not ultimately be thoughts or feelings, or even imaginations and illusions? If more or else than states or acts of mind, what more, what else? If they are affirmed to be existences, or substances, or realities, and the like, what precisely do such affirmations mean? What is truth? Is the assumption that we can attain it well founded or a mere blind belief? If attainable, on what conditions and within what limits is it to be attained? What is knowledge? Is it possible? How is it possible?

How can we separate between the knowable and the unknowable? What are we to think of such assertions as that knowledge is confined to experience, or that spiritual things may be objects of faith but not of knowledge, or that metaphysical problems are incapable of solution?

These are questions with which no special science deals, and which even philosophy as positive does not discuss. Positive philosophy is merely an advance on special science, as special science itself is on ordinary knowledge, and ordinary knowledge on crude and confused sensation. It accepts the sciences and endeavours by their combination and co-ordination to organise knowledge, but it leaves untouched the same questions as the special sciences, and consequently remains as assumptive and dogmatic as they are. For the special sciences and for a consistent positive philosophy, philosophical criticism and philosophical scepticism must be as if they were not. But they undoubtedly exist, and neither can nor ought to be ignored. Philosophy is bound not only to organise but to criticise whatever professes to be knowledge. It must not only survey knowledge as a whole and trace the relations of its parts, but it must satisfy itself as to its grounds and guarantees, and nearly corresponds to what has been designated by Kant *critical philosophy*.

What may be called *positive philosophy* naturally prepares the way for *critical philosophy*. Kant will always be honoured as the man who first adequately realised the necessity under which philosophy lay to exercise its critical functions, and who gave the first general yet profound exposition of philosophy as a criticism of knowledge. He erred seriously, however, even in his conception of its problems, and still more seriously in his attempted solutions. Hence the cry of 'Back to Kant' which for a time resounded widely throughout Germany, and to a considerable extent, although comparatively feebly, in Britain and America, cannot be justly regarded as having been wholly the voice of wisdom.[1] No one, however, has done so much for critical philosophy as Kant. Even his errors have in a wonderful measure proved more valuable than other men's truths.

In the third place, philosophy requires to elaborate a theory of being and becoming in accordance with its views of the sciences and its criticism of knowledge.

Philosophy as critical examines all the assumptions on which philosophy as positive and the special sciences proceed. It is only through critical philo-

[1] See the criticism of Kant's criticism in Hegel's *History of Philosophy*, vol. iii. pp. 423-478 (E.T.); and the author's in *Agnosticism*, pp. 140-190.

sophy that thought can assure itself that what are called science and knowledge have anything to correspond to them,—that their supposition of real objects or objective realities is not a baseless illusion,—that sense and reason are not essentially antagonistic, and experience not inherently self-contradictory. This assurance it may conceivably fail to attain. It may, on the contrary, be forced either to the conclusion that nothing real exists, or that if anything real exists it cannot be known. In other words, its criticism of knowledge may lead to philosophical nihilism or to agnosticism. But it may also issue in the refutation of these hypotheses and the vindication of the beliefs which underlie the special sciences, ordinary knowledge, and common life. It may warrant the conviction that objective reality is the necessary antecedent and universal correlative of the subjective activity in knowledge, and that, so far from being absolutely unknowable, it is continuously self-revealing even to our very limited minds. If this result, however, be reached, philosophy is manifestly bound to endeavour to exhibit the nature of the ultimate reality or realities which the special sciences presuppose and in some measure reveal, but with which they cannot directly deal, first because they are special, and secondly because they are kinds of knowledge, and logically anterior to the criticism of knowledge. Philosophy in this phase—philosophy as the theory of being and becoming—is what has for very long been known as *metaphysical philosophy.*

As such it cannot be satisfied with mere objective appearances or subjective impressions. It must seek to penetrate farther, must seek after the unseen and eternal, and strive if possible to attain some apprehension of ultimate reality, of absolute being, in Nature, Mind, or Deity. Metaphysics has sometimes been identified with Philosophy; but that is to make either the one term or the other useless. Obviously the latter term is the one best entitled to the wider signification. The former, even if considerably restricted, will still be found sufficiently comprehensive for any good purpose. It will appropriately include Ontology, the doctrine of *being* or *reality* as distinct from *phenomenon, appearance,* or *illusion;* Psychology, but *only so far as regards the primary intuitions of reason and their corresponding immutable objects;* and Theology, but *not further than as occupied with Godhead as the one absolute existence.* To a large extent Psychology and Theology are independent of Metaphysics.

The difficulty of defining Metaphysics is well known. I prefer to regard it not as a science but as a function of philosophy, although I do not see any serious objection to such a definition of it as, say, Trendelenburg's,—"the science which considers what is universal in the objects of all the sciences"; and still less to that of Prof. Fraser,—"the knowledge of being in its universal principles." Either knowledge or philosophy seems to me a better generic term for Metaphysics than science. The jocular definition even given of it by De Morgan is decidedly suggestive,—"The science to which ignorance goes to learn its knowledge, and knowledge to learn its ignorance. On which all men agree that it is the key, but no two upon how it is to be put into the lock."

The metaphysical function of philosophy is a most important one. Although it may not be exact science, such science has owed a great deal to it. It has engrossed the attention and energies of many of the world's greatest thinkers. Socrates by his questionings, Plato by his dialogues and dialectic, and Aristotle by the work called (not by himself, however) 'Metaphysics,' were among the first clearly to show what it meant and should aim at accomplishing. The most renowned oriental, medieval, and modern philosophers have been eminent metaphysicians, and their reputations as philosophers have been largely owing to their having been wise enough not to despise

'Metaphysics' rightly understood. There is not the slightest likelihood of 'Metaphysics' becoming extinct. It will be despised only by the foolish or by those who are ignorant of what it means. No highly thoughtful man can fail to be somewhat of a metaphysical cast of mind.

In the fourth place, philosophy ought to forecast, as far as it can, the course of things,—the future of the world and life, of humanity and science,—and to determine what the worth of enjoyment is, and of truth, beauty, virtue, and piety, in relation to one another, and to the great final end of existence.

Philosophy as a science of the sciences, as an inquiry into the nature and limits of knowledge, and as a doctrine of being and becoming,—or, in other words, philosophy as positive, critical, and metaphysical,—is theoretical philosophy in its three stages, and the whole of theoretical philosophy; but not the whole of philosophy, because although philosophy be fundamentally and predominantly theoretical, a merely theoretical philosophy must be essentially incomplete. Practical applicability is a necessary consequence of theoretical accuracy. The true theory of the relations of the sciences, of the conditions of knowledge, and of the nature of existence and causation, must be also the only true basis of doctrine as to the ends and issues, the purposes and destinies of the beings which constitute the universe. Whither tends the physical world? What is the chief end of man? To what goal is society moving? Is life worth living? Is optimism or pessimism or an intermediate hypothesis the legitimate conception of existence? Questions like these can only be answered aright in connection with a general theory of final causes such as a comprehensive and profound philosophy alone can provide. The answers given to them even by the most comprehensive and profound philosophy of the present age, and of many ages to come, may be far from distinct and certain, and yet may gradually approximate to the full truth as time advances and knowledge increases. Philosophy, when engaged in the study of these questions and seeking to be helpful in the guidance of active life, may be appropriately entitled *practical philosophy*.

The four regions of thought now indicated comprise the entire domain of philosophy. Those who would successfully explore that vast domain should begin their investigations with its first region. As I have already indicated, philosophy as positive ought to precede philosophy whether critical or metaphysical or practical. Although the followers of Comte and the advocates of the so-called "scientific philosophy" err greatly in supposing that philosophy is merely the synthesis and generalisation of the positive or special sciences, they are perfectly right in maintaining that philosophy must be based on these sciences, and can only verify itself through accepting and conforming to their conclusions. Philosophy must base itself on the sciences even while searching for their bases. It may conceivably prove science to be illusory, but in doing so it must annihilate itself, as it can only establish its own claim to credence by first vindicating the truthfulness of the sciences and then appealing to their testimony. Thus philosophy as positive must precede philosophy as critical, metaphysical, and practical; and critical philosophy, metaphysical philosophy, and practical philosophy must submit to be attested by the conclusions of a positive philosophy which accepts the well-established results of any and every science.

If the view just stated be approved we shall be freed from the danger of falling into either of two common and hurtful errors. The first is the identification of philosophy with some special science or group of sciences. The narrow notion that one science belongs to philosophy and another not, that the mental sciences are philosophical and the physical sciences non-philosophical, is still prevalent, but is essentially and intensely unphilosophical. There is no objection to using the terms science and philosophy popularly, interchangeably, when no harm is likely to be done thereby; but if we distinguish and delimit them there is but one view of philosophy which can justify itself either historically or logically, and it is that which regards it not as exclusive of any of the sciences, but as comprehensive of them all. From this view it follows immediately, on the one hand, that no special science can claim to be philosophy as against any other special science, and, on the other hand, that no special science is excluded from having the closest connection with and interest in philosophy; that each special science, one may even say each special subject, has its philosophy; the philosophy of any subject as distinguished from the science of that subject being the view or theory of its relations to other things, to the universe of which it is a part, as distinguished from the view or theory of it as isolated or in itself.

The other grave error to which our account of philosophy is directly opposed is that which would found it on common-sense, on ordinary knowledge, on untested and unanalysed consciousness. In pronouncing appeals to common-sense to be illegitimate, I take common-sense in its ordinary acceptation, and censure in no degree appeals to those so-called principles of common-sense which are simply the ultimate conditions of thought as adequately ascertained by psychological analysis. What is alone objectionable is that 'the science of the sciences' should appeal from science to any lower tribunal. Science is more definite and better grounded than ordinary knowledge; nearer the perfect form of human knowledge; such knowledge in its completest and purest state. Therefore whenever science can be had it is with science that philosophy should have to do, and by science that it must be tried and judged. Each science reduces to order, each science develops to perfection or approximate perfection so much ordinary knowledge, and philosophy has to avail itself of the achievements of the separate sciences. Hence an important reduction, an important simplification, of its labour. As far as possible it has to do not directly with the comparative chaos of common knowledge, but with the separate systems of order which constitute the special sciences. Wherever it can do better it ought never to appeal from the higher to the lower tribunal,—from Philip sober to Philip drunk.

III.

Some observations on the various kinds or stages of knowledge still seem to be called for. To apprehend aright the nature of one phase or species of knowledge acquaintance with that of others is indispensable. Yet it has long been, and still is, common to assume that knowledge is only threefold, although the assumption is very erroneous, and has given rise to various false inferences. There are many stages and kinds of knowledge, and so important a fact should not be overlooked or the vast significance of it fail to be realised as fully as possible. Yet there are even scientists and philosophers who treat of ordinary human knowledge as if it were the primary source and oldest form of knowledge. Of course that is a very great error, one which assumes that there was no animal intelligence or knowledge on earth before mankind appeared on its surface, and that the deepest roots of consciousness and thought were brought into the world with the advent of palæolithic man or a primeval Adam. There is not only no warrant for the assumption, but absolutely conclusive evidence to the contrary.

There was animal consciousness on earth for incalculable ages before the genus *homo* appeared on it. Human psychology instead of being the whole of psychology is a very small portion of it. There is a psychology possible of far vaster extent,—a comparative psychology the aim of which should be comprehensive enough to take account of all kinds of creatures that have lived, suffered, and died on earth, and capable of realising aright what their experiences, their inner as well as outer histories, have been. Its task may be a very difficult one, but it cannot be reasonably held to be an impossible one. Why not? Just because man has in his own inmost nature the key to all animal consciousness. In every state of consciousness he has what are called feeling, knowing, and willing, or, in other words, sensation, cognition, and volition. But so has every animal, even the least and meanest. The three elements of consciousness are inseparable alike in man and beast, and hence the former may by a judiciously directed study of the latter acquire a very considerable amount of knowledge of the actions, meanings, and experiences of animals of every kind, and at every stage of their existence. The course of the history of knowledge on earth began apparently with the origination of animal life on earth, although there are some scientists who believe that it began earlier, and that sentiency and consciousness had their roots even in the vegetable kingdom. In proof they have pointed to facts traceable throughout the vegetable kingdom and to adaptations between certain plants and their physical surroundings analogous to those that take place in consequence of the repetition of animal actions and the formation even of human habits. Among the most relevant and best known of such facts are the curious arrangement and action of the leaves in the pitcher-plant; the rapid and peculiar motion which makes the Dionæa muscipula an efficient fly-trap; what Darwin calls the 'nice sense' of the Mimosa; and the elongation and contraction of the stalk of the Vallisneria according as the waters in which it grows rise or fall. But however analogous or akin to animal actions such movements may appear, no one has as yet proved them to be of the same nature, whereas it is certain that knowledge began wherever even the lowest animal life began. All animals have intelligence, and many of them an amazing intelligence. Yet not a few attempts have been made to explain away their intelligence; and to represent their actions as merely automatic, as due to the mechanical play of bodily organs, or to irritability, or to the immediate and sole operations of deity, or to instinct undefined.

It was a curious fact that so late as the year 1874-75 such men as Prof. Huxley (in *The Fortnightly Review*), and Dr Carpenter, Mr Mivart, and the late Duke of Argyll (in *The Contemporary Review*), should have been discussing the question, Whether or not animals are automata? Certainly if animals are automata and their actions automatic so are men and their actions.[1] Man and beasts are alike machines in that they are alike influenced by

[1] See Janet's *Final Causes* (E.T.), Bk. I. c. v., Mechanism and Finality, pp. 137-178.

their physical organisation and alike different from machines in that they are alike endowed with sentiency, appetite, desire, and activity. Sir Isaac Newton, Addison, Bonnet, and others have spoken of the instinctive actions of animals as immediate operations of Deity. They have represented the phenomena of so-called instinct as 'the direct manifestations of the Divine energy in animals,' as 'to be explained by the continued and universal presence of a living intelligent Spirit,' and 'the body of an insect as but a curtain hiding the operations of the Supreme Artist,' — a view which implies that 'God is the soul of brutes,'—an opinion far from peculiarly pious, — a theory which, if consistently carried out, would reduce all nervous actions and all mental processes both in man and beasts to divine operations, and land us in complete pantheism.

Others have represented the study of animal mind as impracticable and futile, on the grounds that we are either (1) not conscious of what takes place in animal mind or (2) that animal consciousness is merely a *quasi-consciousness.* Both reasons are exceedingly weak. If we can know only the mental states of which we are self-conscious it is not merely the minds of beasts that we must remain ignorant of, but every human mind except our own, and also the Divine mind, for all those minds are alike unknown to us except through their self-manifestations to us.

As inconclusive is it to assert that animal consciousness cannot be apprehended and interpreted by us because it is only *quasi-consciousness.* The reply to that is obvious. How can what is asserted as known be known by those who assert it? To be entitled to say what they do say they must have already done what they declare cannot be done; must have interpreted animal consciousness and ascertained what it is before they can rationally believe or pronounce it to be anything—or even *quasi* anything—else. Unless they know what it is, how do they know that it is not such consciousness as they themselves possess, but a mysterious *tertium quid* between that consciousness and unconsciousness. As to the second reason referred to, a *quasi-consciousness* is an absurdity. To call the pain which an animal gives evidence of suffering *quasi-pain* should be recognised by every sane person as an abuse of language. There is no *medium, tertium quid,* or *quasi* in such a case. There is either pain or not pain, sensation or non-sensation, knowledge or ignorance.

Seeing that consciousness and knowledge belong to all creatures in the animal kingdom, man as the earthly head of that kingdom is not only self-conscious and self-cognitive but capable of understanding what are the sentiency and knowledge of the countless active beings which Zoology distributes into its multitudinous divisions,—its types, classes, orders, sub-orders, and families. The psychical life and consciousness of all mere animals is much simpler and more limited than that of man, and may naturally be found, in consequence, to be much more easily understood. That animal intelligence is, as a whole, however, a lower stage of intelligence than the human, and that in every animal species the variation is greater than in the human, must be admitted, and the main reason for such being the case seems obviously to be that the animal mind is much more dependent on the bodily organism than the human mind is on the human body. The former is in comparison much less free. Whereas the manifestations of knowledge in animals are often seemingly automatic, in man they are, in comparison, very exceptionally so. Were it otherwise, the achievements of many species of animals would be far more extraordinary than those of a similar character performed by man. Some of the smallest species of animals display the largest amount of intelligence. The elephant is sagacious within certain limits, and in comparison with the rhinoceros or hippopotamus, but its knowledge is far less wonderful and exact than the knowledge of ants, bees, and beavers. Ants are not only capable of giving good lessons to sluggards but they display a marvellous knowledge in an architecture of their own, the importance of social organisation, and how to conduct war with vast numbers. Bees in the construction of the cells of their honeycombs not only solved an economic problem of the utmost practical importance to them, but which was also so difficult a problem of the higher mathematics that a completely satisfactory solution was first given by Colin Maclaurin in the *Transactions of the Royal Society of London.* The naturalists who have made a special study of the operations and habits of beavers are agreed as to their extraordinary intelligence. In one well-authenticated case these creatures have been proved to have, for generation after generation during at least a thousand years, constructed their lodges, dams, and canals, so as to have at length changed the entire configuration of the region in which they had operated.

The whole animal world is participant in knowledge. Every kind of living creature has some measure of intelligence, sentiency, and self-activity. Whence come they? Whence has every living creature its share of them? Surely not from mere matter in any form, nor from the creatures themselves by any self-creative power, but only from an eternal self-existent Intelligence, an Intelligence to which no origin or limit can be assigned, an infinite and ever-living Being creative and comprehensive of all that knows and all that is known. Apparently there were many and long ages before there was any life and intelligence on earth; but conceivably also the sources of consciousness and knowledge may have been present in the cosmic ether before our world became a globe differentiated from all other worlds. Nor is it entirely certain, perhaps, that vegetable and animal vitality may not have had in an incalculably remote age on earth their origins in the same protoplasmic substance. What is alone indubitable is that conscious life has had an exceedingly long history on earth.

That it was preceded by a vastly long history of entirely dead matter does not seem to have been adequately proved either in the affirmative or negative. Even a molecule of matter would appear to have a history in or behind which alike the chemist and biologist, geologist and palæontologist, have failed entirely to decipher. No educated person, however, thanks to the labours of those scientists, can now fail to believe that the history of animal life and intelligence has already been one of amazing and incalculable length, as well as vast breadth, and that from its first known appearance until the present time it has been a history of unbroken continuity the development of which can be traced as plainly as the history of any individual. It has passed through many epochs and phases, every epoch having its own physiognomy and every phase of every epoch having presented some variation, but there has nowhere been complete separation of the stages or radical difference between the species of animals that have lived and worked, enjoyed and suffered in those stages. Absolutely new and original species, however, have been nowhere discovered. From the earliest time animal nature has had general features in common with those of to-day. It is impossible to draw an absolute limit between the beings that have existed before us and those that are living around us.

Our animal world is not distinct from the fossil world, but rests on it and is the continuation of it at almost every point. The two in alliance have had a series and history of the stages which are so many periods of progress alike in the general history of the animal world and in the special history of mankind. And hence there has been in the main a continuous growth of animal and human intelligence and knowledge towards development and improvement. The numbers of animals and men have been increased. There has been greater differentiation alike of their physical and mental organisation. There has likewise been progress as regards sensibility, intelligence, and activity—*e.g.*, improvement in sight, hearing, smell, affectionate and social sentiments, a higher development of the nervous system, and more combination and co-operation.

One has a temptation to dwell on so interesting a subject, but I must not yield to it, as it has been in recent times dwelt on by many distinguished scientists. The study of animal mind had been inaugurated by Aristotle's *History of Animals*, yet during the last fifty or sixty years it will scarcely be questioned to have been more carefully and fruitfully cultivated than all those which had preceded them. Comparative Psychology is mainly the creation of the present age, during which there has, perhaps, been no more interesting scientific achievement. It has immensely extended the sphere of psychological study. Among those who deserve most credit for that result have been Bingley, Büchner, Darwin, Gaudry, Houzeau, Huber, Jesse, Lubbock, Perty, Romanes, Semper, and Wundt. They are all authors of most instructive and easily procurable works.

Of human knowledge there are universally recognised to be three kinds or stages—viz., ordinary, scientific, and philosophic knowledge. Ordinary knowledge is the kind of knowledge common to all sane men but also such knowledge as is often extremely indistinct, confused, and superficial. It is not strictly definable and generally very vague as to its contents. The nature of it is common not to men only but to all animals. It is distinguishable from science by its lack of precision and exactness and from philosophy by its lack of comprehensiveness and profundity. Even as knowledge of particular objects and limited ends indeed it implies universal principles and rational intuitions but is not consciously and distinctly aware of them. Only in the scientific and philosophical stages do they come clearly to light. Yet ordinary knowledge is a knowledge by no means to be despised. A large portion of it is probably of more value than much which is called science and believed in as such. Although less exact than science it is often less capable of being dispensed with. A human world composed exclusively of scientific experts might very possibly, and not very improbably, be not better but worse than one like the present composed for the most part of merely ordinarily intelligent men. There is a vast amount of ordinary knowledge which is more helpful and of more real human interest than there is of science. All the roots of scientific thinking are already in ordinary knowledge. Compared with ordinary thought the amount of scientific thought is very limited.

Scientific knowledge is, nevertheless, rightly regarded as on the whole a higher stage of knowledge than non-scientific or ordinary knowledge. It is a knowledge of more than mere facts or common observations and experiences, including as it does a search for the reasons and causes of things as well as of mere perceptions of them, or, in Greek phraseology, not merely the ὅτι but also the διότι of phenomena. And, further, all scientific knowledge is knowledge of a specific kind, and differentiated from knowledge not of that kind. Each science has a sphere of its own, and is not to be confounded with ununified and indeterminate knowledge. The scientist is a specialist, and as such one who keeps within a province peculiarly his own, and distinguishes it from other provinces, although if a wise man he will look beyond it and take note of what other scientists are doing in contiguous departments. The methods appropriate to the several sciences must vary with their objects. Still less, of course, is scientific knowledge to be identified with mere belief, or with mere art and practice, than with ordinary knowledge. To collect facts, to analyse material objects and mental states, to distinguish between semblance and reality, to discover and formulate laws of sequence, to bring to light the conditions of order and organisation alike in the physical and spiritual worlds, are what the sciences have to accomplish, each in its own province.

Science is often, but not always, what had been merely ordinary knowledge in an advanced and improved stage. It is not always so because when sciences are thoroughly established they are often capable of evolution from within so as to yield vast accessions to knowledge such as have never existed except in scientific form. Mathematics is constantly thus extending itself into regions where unscientific intellect has never been, and consequently can predict effects which have never been observed, and may be carried to developments far beyond the reach of experiments. But in general science issues out of ordinary knowledge, and that knowledge may in every case be regarded as a step towards scientific knowledge,—as a humbler stage always, a prior stage generally of the same movement or process. Science rises superior to ordinary knowledge in being both more general and more definite. More general inasmuch as it regards things not as isolated and individual but as included under some law, as terms of some fixed relation, of coexistence, or succession; and more definite as implying a recognition of the *exact* relation in which one fact stands to another, whereas ordinary knowledge in its recognition of connection between facts is merely of *some sort* of connection.

In a general way philosophic knowledge may reasonably be held to be the highest stage and most comprehensive kind of human knowledge, but only when it strives with a fair measure of success to realise the ideal at which it aims. All that assumes to be philosophy is not to be taken simply on its own authority. Much of it has been found to be instead of perfect knowledge pretentious nonsense. But genuine philosophy is worthy of all the praise which has been bestowed upon it. Wherever there has been active and earnest thinking, wherever the arts have flourished, wherever the sciences have prospered, wherever civilisation has spread, there philosophy can be shown to have been at work. The term itself and the history of it have been suggestive and instructive as to what it has meant and ought to mean. It was as "the love of wisdom," and not as the acquisition of mere knowledge, that it was called into existence, and the Pythagoreans and Platonists continued to regard "the yearning after divine wisdom" as what was properly distinctive of it. Cicero spoke of it as "the science of divine and human things and of the causes in which they are contained." Descartes changed, and contributed to modernise, the conception of it, by representing it as "the pursuit of the perfect knowledge of all things that men can know, deduced from first principles." Kant described it as "the science of the relations of all knowledge to the essential ends of human reason." Lotze's definition of it (in his *Grundzüge der Logik*, § 83) is "an effort to import unity and connectedness into the scattered doctrines of cultured thought, to follow each of these directions into its assumptions and into its consequences, to combine them together, to remove their contradictions, and to form out of them a comprehensive view of the world; mainly, however, to subject the ideas which science and life regard as principles to a special scrutiny in order to determine the limits of their validity." Even those few definitions may suffice to show what has been the course of thought as to the nature of philosophy. It has been a long course and one never entirely interrupted. Philosophy has always preceded what we would call science. Wherever there is earnest human thought as to truth and error, good and evil, right and wrong, there is something of the nature of philosophy, and as such it aspires to be coextensive with human knowledge, claims the right of criticising and testing all opinions, and hesitates not to raise and try to answer the most difficult and perplexing yet engrossing and important questions which can come before the human mind. Hence philosophy is rightly, and almost universally, regarded as the last and highest stage of human intelligence.

Philosophy, in order to be as comprehensive as it ought, has to deal as its subject with the entire intelligible universe, the three final existences of which are God, the world, and self. Its ways or modes of manifestation and action are:—1°, Positive or Phenomenological; 2°, Critical or Epistemological; 3°, Metaphysical or Theoretical; and 4°, Practical; or, it may suffice to say simply the positive, the critical, the metaphysical, and the practical.

Philosophy as universal science has, in the first place, to deal in a comprehensive and general way with what all the special *positive sciences* deal with in a sectional way. It has to seek to attain to a knowledge of the unity, self-consistency, and harmony of the teachings of these separate sciences, and to a knowledge of what the universe is according to their collective testimony. Philosophy as thus a synthesis of the positive sciences is Positive Philosophy. As such it deals only with phenomena, appearances, particular experiences,—with what the ordinary man and the positivist scientist accept as alone facts. According to Comte and the adherents of all the positivist schools there is no other philosophy than such *positive* philosophy. In that they err, but they are in no way mistaken in maintaining that there is a positive philosophy, and that it is of primary and fundamental importance. They are only mistaken in supposing that philosophy can rationally stop where they would have it to do.

Philosophy should be *critical* as well as *positive*. A merely positive philosophy must be a very imperfect philosophy. Philosophy as positive is far from an adequate ideal of philosophy. Even scientific thought is not necessarily self-criticising thought; on the contrary, mere scientific thought, however rigid and methodical, is essentially dogmatic thought,—reasoned yet unreflective thought. It builds up what is admitted to be knowledge, but it does not inquire what so-called knowledge is or is essentially worth. The mere scientist often fancies that he is a man who takes nothing on trust, when, in reality, he is taking everything on trust, because he accepts without question or reservation thought itself as naturally truthful, and its laws as valid. Whatever superficial scientists may suppose to the contrary, the fact is that the entire procedure of science, and of philosophy in so far as it is merely a generalisation of science, is assumptive and dogmatic. The science which is so often contrasted and opposed to faith by sceptics is frequently implicit faith, and in the view of a serious and consistent scepticism must be deemed a blind faith. Thought may assume, however, and is even bound to assume, a very different attitude towards itself and its own objects; not only may but ought to pass from a believing to an inquiring, from a dogmatic to a critical stage, from a study merely of the superficial and apparent in knowledge to an examination of the conditions and guarantees of knowledge. Philosophy, in a word, has not only to accumulate what passes for knowledge in the opinion of positivists, but must assure itself as to the solidity of its own foundations. As critical it is occupied with a fundamental and universal problem, the problem as to the possibility and reality of knowledge of every kind, if philosophy is not to end in nihilism or agnosticism. It is essentially epistemology (inclusive of what is philosophical in logic and methodology).

Philosophy, besides being positive and critical, should also be *metaphysical* (systematical or theoretical). The criticism of what passes for knowledge may lead only to a negative or sceptical result, either to philosophical nihilism or agnosticism. Were it to be successful, however, all so-called science must be but an inevitable and ineradicable illusion, and all so-called knowledge at bottom no knowledge, or the knowledge of nothing. In that case philosophy might be best defined as a demonstration of the vanity of thought. While, however, its criticism of knowledge may conceivably lead to philosophical nihilism or agnosticism, it may also, on the contrary, issue in the refutation of them and the vindication of the beliefs which underlie the special sciences, ordinary knowledge, and common life. In other words, it may warrant the conviction that objective reality is the necessary antecedent and universal correlative of the subjective activity in knowledge, and so far from being absolutely unknowable is continuously revealing itself, even to our very limited minds. But if that result be reached, philosophy is manifestly bound to exhibit the nature of the ultimate which the special sciences presuppose and so far manifest, but with which they cannot competently deal—first, because they are special, and, secondly, because they are logically anterior to the criticism of knowledge. Philosophy in that phase has for very long been known as metaphysical or ontological philosophy. It has also been often termed systematic, theoretic, or speculative. Of course, philosophy as metaphysical has to determine whether or not there is God, the ground and source of all being, the reason of all existences and events, and cannot escape the necessity of being either theistic or antitheistic. It has to deal with all dogmatic metaphysical theories, and all such theories must be either *theistic or anti-theistic.*

Hence it cannot itself escape the necessity of being either theistic or anti-theistic. If the former be arrived at, it is its obvious duty to tell us what it can of God, of the world in relation to God, of man in relation to God, of providence and theodicy, of revelation and its media, of the destination of mankind and the consummation of things, of the aims, ideals, spheres of action of the religious life, and the like. A theistic metaphysical philosophy is bound in self-consistency to exhibit the knowledge of God as the alone absolute and all-comprehensive knowledge,—the idea of ideas in metaphysical language, —and as inclusive of all the categories of being and thought in their perfection. A correct doctrine of the nature and function of the categories in thought shows what is meant by knowing God as the absolute, why it is erroneous to say that we cannot know God, seeing that we can only know the relative or the phenomenal, and the categories are only valid for experience. In reality, all progress in speculation, in science, in moral experience, and in spiritual life, promotes progress in knowledge of God.

Philosophy as a scheme of the sciences, as an inquiry into the nature and limits of knowledge, and as a doctrine of being and becoming,—or, in other words, philosophy as positive, critical, and metaphysical,—is theoretical philosophy in its three stages, and the whole of theoretical philosophy. It is not the whole of philosophy, however, because although philosophy be fundamentally and predominantly theoretical, a merely theoretical philosophy must be essentially incomplete. Practical applicability is a necessary consequence of theoretical accuracy. The true theory of the relations of the sciences, of the conditions of knowledge, and of the nature of existence and causation must be also the only true basis of doctrine as to the ends and issues, the purposes and destinies, of the beings which constitute the universe. Whither tends the physical world? What is the chief end of man? To what goal is society moving? Is life worth living? Is optimism or pessimism or an intermediate hypothesis the legitimate conception of existence? Questions like these can be answered aright only in connection with a general theory of final causes such as a comprehensive and profound philosophy can alone provide. The answers given to them even by the most comprehensive and profound philosophy of the present age, and of many ages to come, may be far from distinct and certain, and yet may gradually approximate to the full truth as time advances and knowledge increases. Philosophy when engaged in the study of these questions and seeking to be helpful in the guidance of active life may be appropriately entitled *practical philosophy.*

Philosophy may not unreasonably present a claim to be regarded as the highest and most comprehensive kind of all human knowledge but certainly not of all knowledge. There is an infinitely vaster and more perfect knowledge than any to which man or any other or even all created beings can pretend to possess. There is a knowledge which we are very apt to ignore although all other knowledge in the universe springs from it and is closely connected with it. In other words, there are not merely ordinary and human knowledge, science, and philosophy, but *omniscience*, — divine intelligence and wisdom, — an all-comprehensive, perfect, and infinite knowledge. Nothing can be hid from God. All is perfectly known to Him in the past, present, and future, from the highest to the lowest, and from the least to the greatest. He has all the perfections of knowledge in himself and also knows all that there is to know from without. Co-extensive with omniscience is omnipotence. They are indissolubly united. The former is not inactive nor the latter unenlightened. More need not here be said. The subject has been treated of in every comprehensive system of theology.

A HISTORY OF CLASSIFICATIONS OF THE SCIENCES

E

A HISTORY OF CLASSIFICATIONS OF THE SCIENCES.

THE first problem with which philosophy, alike as *scientia scientiarum* and as *positive philosophy*, should deal seems to be how may the sciences be rationally arranged and classified. Unless it be so far accomplished obviously no attempt at the organisation of either knowledge or science can be successful. Philosophers have always felt, more or less distinctly, that such must be the case. They have never shown themselves wholly unconscious that they ought to aim at the organisation of knowledge. On the contrary, they have made many endeavours to realise that aim, and have always recognised that the first step or stage to the organisation required is some form of classification. From the time of Plato to the present day there has been a continuous series of attempts to classify the sciences. An historical and critical account of them can hardly fail to be useful, even although none of them may have been completely successful. Indeed, no scheme of the sciences can be final and perfect so long as new sciences remain to be formed. On the other hand, few have been entirely worthless, and some may fairly be held to have been of much value. A study of them is indispensable at least to those who would improve on them. It is always helpful towards knowing how a thing ought to be done to consider how it has been done. Thus only can all the points of view, principles, and methods which require to be considered in connection with any difficult problem be brought distinctly before us.

I. FROM PLATO TO THE RENAISSANCE.

Platonic scheme.

Plato was, perhaps, the first who sought to give a systematic distribution of knowledge. We must be careful not to confound with that distribution so-called divisions of his philosophy. Of its very nature his philosophy will not divide. Those who have divided it, like Marbach into general and applied, or, like Krug into theoretical and practical, have overlooked the fact, which numerous passages might be brought to substantiate, that, in the eyes of Plato, philosophy was an essentially practical spiritual process. It was not theory or practice, science or life, nor even theory applied to practice, science applied to life, but both in one—the striving of the soul to purify and ennoble itself, to make itself all beautiful within. There is no right understanding of the philosophy of Plato possible if we forget that he regarded it as primarily a process, the true life of the spirit, the soul making of itself a divine poem, the highest music. It is equally incorrect to divide the Platonic philosophy, as Van Heusde has done, into a philosophy of the true, of the beautiful, and of the good. That is an altogether modern mode of dividing philosophy, and quite contrary to the spirit of Platonism. Philosophy was, according to Plato, not only essentially practical, but also essentially one, and one because all ideas lead up to the idea of the good.

The division of philosophy most commonly attributed to him, however, is that into dialectics, physics, and ethics. But although Schwegler, Zeller, Ferrier, Ueberweg, Erdmann, and many others, have adopted it as substantially warranted, it can exhibit no valid claims. It is admitted that Plato nowhere distinctly states it. The very names physics and ethics are unknown to him, and dialectics is with him not a part of philosophy, but the whole of philosophy. The way in which he came to be credited with the division is apparent from what Sextus Empiricus, who flourished about the beginning of the third century, says on the subject: "Of those who divide philosophy into physics, ethics, and logic, Plato is *virtually the originator* (δυνάμει ἀρχηγός), having discoursed on many physical, many ethical, and not a few logical questions."[1] The latter clause is here obviously the explanation and reason of the former. It is because Plato has discoursed much on physical matters, much on ethical matters, and not a little on logic that he is affirmed to have been virtually the author of the threefold division of philosophy which was afterwards widely prevalent. There is, in fact, no other ground on which it can be carried up to Plato with any plausibility, and this ground is quite insufficient. That Plato wrote on all these three subjects cannot in any degree warrant us to call him even the virtual originator of the distribution. It was scarcely possible that Plato, or any other person, should write much on philosophy without handling to some extent both physics and ethics, and wholly impossible to handle them without keeping them in some measure apart, but that was a very different thing from making physics and ethics distinct parts of philosophy, co-ordinate with each other and with dialectics. That Plato certainly did not. There is no dialogue of Plato exclusively

[1] *Adv. Math.*, vii. 16.

dialectical, and no dialogue from which dialectics is excluded. Physical, ethical, and all other inquiries are only included in his philosophy in so far as they are dialectical, and his dialectics exists only as dealing with the idealities of nature and spirit. Plato knows nothing of a logic which has a province of its own apart from all definite ideal contents. It is vain to try to classify his writings as dialectical, physical, and ethical.

Plato's distribution of knowledge is one involved in his very theory of knowledge. It has been discussed so often that I shall treat of it as briefly as possible, and only because I must. According to Plato, then, two worlds lie before the vision of man, —a visible world and an intelligible world. Each of these worlds, in its turn, divides into two. Thus the visible world is made up either of things or of images of things. The former are the rocks, trees, animals, &c.; the latter are the shadows and reflections which they throw off, — such shadows and reflections as may be seen in water or in a mirror. All the objects of the visible world are discerned only through sense (αἴσθησις), but sense in contact with things generates belief (πίστις), while in contact with images (εἰκόνες) it generates merely conjecture (εἰκασία). Belief and conjecture are but a higher and lower form of opinion (δόξα). Belief differs from conjecture; views based on things are not to be confounded with views based on mere shadows, and have a greater worth and usefulness; but in no form can the informations of sense give us truth or be entitled to the name of knowledge.

There is, however, an intelligible world, with objects which reason apprehends and not sense. These objects are likewise divisible into two classes, —conceptions and ideas. Conceptions are on the lower level, and the mind reaches them by the help of certain objects of sense which are a sort of images of them. The mathematical sciences are conversant with them, and in these sciences we make use of visible figures, and motions, and audible sounds, but only to help us to the comprehension of forms, properties, and ratios, which intellect alone can grasp. They are five in number, and form a naturally and closely connected series,—Arithmetic, Plane Geometry, Solid Geometry, Astronomy, and Harmonics. Even the two latter deal not with physical things,—the visible luminaries of the sky, and the musical sounds of the voice and other instruments,—but with permanent truths, mathematical relations, which eye cannot see nor ear hear.

Plato gives, in the seventh book of the Republic, a very remarkable account of the sciences conversant with conceptions. To that account it must suffice here merely to refer. The great value of those sciences in his view was that they tended to raise the mind above themselves, to develop philosophic insight, to educate reason to apprehend the absolute truth which is the light and life of the soul. In themselves he regarded them as inherently defective. They begin with certain assumptions and give us only the consequences which follow from reasoning on these assumptions. They start from principles which they cannot prove, which it is beyond their province to prove. They are essentially hypothetical. There is need, accordingly, for a higher science; a science which may make use of the assumptions of the sciences which deal with conceptions as occasions and starting-points whence it may ascend to absolute principles, to what has its reality and evidence in itself, to ideas. And there is such a science. Its name is Dialectic. The lower sciences have for their objects conceptions or scientific assumptions; the faculty which they employ is discursive reason, and their procedure is demonstration. The highest science has for its objects ideas, not conceptions; absolute, not hypothetical principles; real, not assumed existences; for its process intuition, not demonstration, and for its faculty the intuitive, not the discursive reason. It includes in itself all properly philosophical investigations. It is at once a metaphysics, a logic, a theology, an ethics, and an æsthetics; a metaphysics, because occupied with the immutable and invisible; a logic, because the form and method of absolute science; a theology, because the supreme idea is the ultimate cause; an ethics, inasmuch as conversant with the principles which are the source of all morality; and an æsthetics, since true beauty is ideal and transcendental in nature and origin.

Plato's doctrine of science originated in a profound conception of the nature of intelligence, and corresponded to a magnificent view of the universe of existence. From its promulgation to the present time it has captivated alike the reason, imaginations, and moral susceptibilities of men as no similar theory has done. But, whatever were its merits, it had also defects, which showed themselves very plainly in the Platonic survey of the sciences, and which led, in particular, to undue contraction of the sphere of science. The whole world of sense is not to be relegated, as Plato advised, to the limbo of mere opinion. Natural apprehension and ordinary judgment are not so essentially different from scientific cognition as he assumed. The notion that there is no science of phenomena, and that consequently science cannot be reached through the study of phenomena, but requires us to get beyond phenomena, through and above them as it were, into a region of types, exemplars, conceptions, ideas, is directly antagonistic to the spirit of modern science, and has been amply confuted by the splendid achievements of modern science. It is a notion which involves denial of the possibility of the physical or natural sciences.

It apparently led Plato to that conclusion. For although in the Timæus he speculated on the origin and disposition of the world, and the organisation of man, he expressly held that nothing could be affirmed on these subjects as certain. What is called his Physics was an application of his Dialectics, and of a character which he himself maintained must be conjectural. Of physical science in the proper sense he has shown, I think, no conception. The error which led him thus unduly to restrict the sphere of science he also carried into his actual survey and description of the sciences. There it took the form of the dogma that the realities of a science are distinct from its phenomena. The latter do not contain or manifest, but only suggest the truths of science, and aid the mind to reach them. The conceptions of Geometry are ideal assumptions; its phenomena are visible illustrations which never exactly correspond to them, and often do them great injustice. So there is an Astronomy of theories or realities, and an Astronomy of appearances or phenomena; and the latter is not true Astronomy, because the variegated adornments which appear in the sky, the visible luminaries, beautiful as they are, are only a sort of admirable diagrams by the help of which we may rise to the contemplation of spheres, movements, and relations, which are real and immutable, and which can be grasped only in mental conception.

Now, all that is untenable. The diagrams of the geometer are not phenomena of geometry. Geometrical reasoning refers entirely to ideal figures and relations, understanding thereby immediately or mediately defined figures in immediately or mediately defined relations. However badly drawn may be the diagrams before the bodily eye of the geometer, those before his mental eye are always absolutely accurate delineations. He can only reason on the supposition that his triangles, squares, &c., are precisely what they are defined to be. It is likewise vain to separate and contrast an astronomy of appearances and an astronomy of theories. The appearances are in astronomy the very things and the only things to be explained. A theory, to be of any worth, must be one which accounts for the appearances. Plato failed to perceive how phenomena exhibit laws and how laws manifest themselves in phenomena, and consequently he opposed phenomena to realities in a way which few will now undertake to defend.

Apart from the error indicated, Plato's survey of the hypothetical sciences—the sciences which deal with conceptions—is of remarkable merit, considering the age to which it belongs. It is especially meritorious for the comprehensiveness and correctness of mathematical view which it displays. It strongly corroborates the historical testimony that Plato was a proficient in the mathematical knowledge of his time. He ignored, as we have seen, natural science. Here, where Aristotle was so powerful, Plato was comparatively feeble; but, on the other hand, where Aristotle was weakest Plato was strongest. All the difficulties which intelligence meets with may be reduced to two classes,—difficulties of abstraction and difficulties of complexity. Of superior minds some overcome more easily the one class of difficulties and some the other class. Aristotle was the more fitted to deal with the complex, Plato with the abstract. Hence, Aristotle was drawn to natural philosophy, and still more to natural history and psychology, and whatever demanded close observation and searching analysis; Plato to mathematics, and all those loftier problems which most transcend sense and most exercise pure intellect. Few thinkers have discerned so broadly and clearly as Plato the relations of the mathematical sciences to philosophy.

Aristotelian scheme.

Aristotle's conception of philosophy as distinguished from science was greatly inferior to that of Plato, and his criticism of the nature of knowledge was far less profound and suggestive, yet his work was, on the whole, an advance on that of his predecessor. It was at once its continuation and complement. Aristotle collected the truths which Plato had so lavishly scattered, added to them a multitude of facts acquired by his own indefatigable industry, and a multitude of reflections suggested by his own vigorous and penetrating intellect, and combined with rare judgment his vast acquisitions into distinct organic systems. He thus became the founder of more sciences than any other man. He gave existence and form to almost as many special scientific disciplines as he wrote books.

That great thinker, than whom there probably never lived a man of more encyclopædic mind, adopted a threefold division of philosophy, science, or knowledge. He distributed it into Theoretic, Productive, and Practical. Theoretic Philosophy has no aim beyond the apprehension of truth. It is conversant with the existent, with being. It subdivides into Physics, Mathematics, and Metaphysics. Being, considered in connection with whatever can be known through perception and experience, is the subject-matter of Physics, which, according to Aristotle, includes Psychology. Being, conceived of apart from the variations of the material world, but not apart from matter, is that with which Mathematics is conversant. Mathematics consequently differs from Physics not essentially, but only in degree, as being more general and abstract. Metaphysics, again, differs from Mathematics just as Mathematics differs from Physics, being still more general and abstract. It treats of Being *per se*, of the existent in its absolute nature and universal properties. Aristotle called it "First Philosophy," and sometimes "Theology." It contained what little theology he taught.

But philosophy, according to Aristotle, although primarily is not exclusively theoretic. The contemplation of being is its proper function in its purest form, but not its only function. It has regard also to the production of effects and to the regulation of human actions. In the former case it is Productive Philosophy; in the latter case it is Practical Philosophy. Productive Philosophy differs from Theoretic Philosophy because it tends to performance instead of to contemplation, and from Practical Philosophy because it does not terminate in the regulation of actions, but in the origination of permanent products. It is the theory of the arts. Aristotle did not subdivide it. His "Poetics" deals only with one of the "imitative" arts. Rhetoric, which, judging from its general character, one expects to find placed by the side of Poetics, was viewed by him as a science auxiliary to Politics. Practical Philosophy looks beyond truth to the good, and seeks so to regulate actions that the good may be reached. Its two chief branches are Ethics and Politics. The former deals with man in relation to his natural good as an individual; the latter is an inquiry as to how society should be constituted with a view to the public good.

Within this scheme Aristotle did not place Analytics, later called Logic. He regarded it not as a part of philosophy, but as an introduction to philosophy, and especially to "first philosophy." As a doctrine of the principles and processes of science he considered that it ought to take precedence of the sciences. This, of course, was virtually to exclude it from the sciences and to allow that the proposed classification of the sciences was not inclusive of all departments of knowledge, while it could, with much appearance at least of truth, be maintained that the principles and processes of science are only ascertainable after sciences have been formed. Logic may, however, have a place assigned it within the Aristotelian scheme. It may, indeed, be ranked either among the Productive or the Practical Sciences; among the former if its end be supposed to be the production of arguments; among the latter if it be held to aim at the regulation of the reasoning faculty. Rhetoric, also, is virtually excluded from the classification when represented as simply an auxiliary to Politics. It too, however, like Logic, may easily be placed within it, and either as a Productive or a Practical Science,—the former if its aim be deemed the production of orations, the latter if it be regarded as looking to influence on the mind and conduct. Economics is conjoined by Aristotle with Rhetoric, as being also a science auxiliary to Politics. It might just as well be viewed as a constituent member of the group of Practical Sciences.

The work which Aristotle accomplished in the way of originating and advancing the sciences which he arranged or classified according to the plan now described, gained him a unique position in the history of science. No one has attained, or can reasonably hope to attain, any very like position. The scheme of classification itself, however, has obvious defects. Thus, in the first place, the distinction between Productive Sciences and Practical Sciences ought not to have the importance which is assigned to it. It is neither broad nor deep, and certainly not fundamental or primary. Nay, it is much to be doubted whether it is a distinction which can be at all applied to separate and distribute the sciences. For as every science is in some measure both regulative of actions and productive of results, it would seem that there must be arbitrariness in forming sciences into groups by viewing some sciences as only regulative of actions and others as only productive of results. Aristotle chose to regard Politics, for example, as a Practical Science, but he might with equal reason have ranked it as a Productive Science. He might have looked at the result it seeks to accomplish rather than at its character as a means, and the result is a permanent product,—an orderly, prosperous, and enduring society.

In the second place, it is erroneous to classify the sciences according to ends, either of regulation or production. They should be arranged according to their natures, their inherent characteristics, not according to anything lying beyond themselves. The end of a science is not anything fixed. It is the sum of the uses to which the science can be put, and uses always vary with wants. One science may have many ends, and many sciences may require to be combined in order to gain one end. It must be especially erroneous to arrange some sciences according to their natures and others according to their ends. It must be illegitimate to employ two principles of classification, and when one fails, to have recourse to the other. That is a procedure which must at once give rise to cross-divisions, and which has in itself no logical limits. If we can introduce two principles, why not three? And if three, why not as many as there are things to divide? There can be no legitimate scheme of classification in which the divisions are not determined throughout by one common principle.

That leads me to say that a third, and perhaps the greatest, defect of the Aristotelian survey of the sciences was the want of unity which arose from the absence of a philosophy inclusive of, but superior to, the sciences. Without explicitly affirming that he did so, he, in reality, viewed philosophy as merely a whole constituted by the sciences, a sum made up of the sciences as a unit is made up of its component fractions. But this leaves no philosophy distinct from the sciences, and either able or entitled to co-ordinate and organise them. Hence in the Aristotelian arrangement there is a certain grouping of the sciences, but not a real systemisation of them. They are not shown to constitute an organic whole. They have each an independent foundation, and they are also in some degree classified, but there is no highest science to comprehend them and to determine the place of each. What Aristotle called First Philosophy and his commentators Metaphysics, does not perform this function. Its object is being as being, and so it is the antecedent and presupposition of all other sciences, since they all treat of special concrete beings, but it possesses a merely abstract universality, and it has no power nor is it any part of its business to organise the various sciences into a system. It is not, to use an Aristotelian word, an architectonic science. The science which Aristotle himself regards as such is Politics, but its claims to the honour are altogether inadmissible. They amount merely to an affirmation that Politics is entitled to control other sciences, seeing that politicians must view the sciences in relation to the public good. We may be sure, however, that the order of the sciences has a far deeper source than the will and the interest of men. It must spring from the essential truth of things, from the all-pervasive order of nature.

Modified Aristotelian classification.

The Aristotelian classification, notwithstanding its radical defects, was widely accepted, although only in a slightly modified form. The narrow, the really untenable distinction between Productive and Practical Sciences was dropped, and philosophy came to be divided simply into two great branches, the Theoretical and Practical. This division found recognition both among the Stoics and the Epicureans. Some expressed it by representing Philosophy as either Physical or Ethical, *i.e.*, either concerned with the contemplation of nature or the regulation of human action. The great objection to it is that it identified, or rather confounded, philosophy with science. It recognised no philosophy distinct from the sciences. It assumed that the branches of philosophy were the divisions of the sciences. If that be the case there is either no philosophy proper or no science proper, for either philosophy absorbs the sciences or the sciences leave no room for philosophy. The division of the sciences into Theoretical and Practical is still a favourite popular one. There can be little doubt, however, that it is faulty, even when science and philosophy are expressly distinguished. All the so-called Theoretical Sciences may be regarded as also Practical Sciences, and all the Practical Sciences as also Theoretical Sciences, if each class be only looked at from the point of view previously appropriated to the other.

The division of philosophy into Dialectics, Physics, and Ethics, commonly but erroneously attributed to Plato, has been also referred to Aristotle, although it is, of course, admitted not to have been the one which he himself adopted. It has been referred to, however, on the authority of a passage which by no means warrants the conclusion drawn from it. In that passage (*Topics*, B. I. ch. xiv.) he says that "there are three parts of propositions and of problems; for some propositions are ethical, others physical, and others logical"; and he says so only when treating of the choice of propositions with reference to disputation. To regard that as a division of philosophy into Physics, Ethics, and Logic is to raise a very large superstructure on a very small foundation. To classify propositions with reference to a particular end is a very different thing from classifying the sciences. Besides, Aristotle put forth his classification of propositions as only generally, only in outline, true (ὡς τύπῳ περιλαβεῖν).

Stoic and Epicurean distribution of sciences.

The threefold division of philosophy into Logic, Physics, and Ethics can be fairly ascribed neither to Plato or Aristotle. It may have been enunciated by Xenocrates, as Sextus Empiricus says, but there is now no proof of that, and not unlikely it originated with those who attached so much importance to it, the Stoics. They regarded all knowledge as vain and superfluous which had no end beyond itself,—which did not help towards the attainment of that wisdom to which the character and conduct ought to conform. They held that philosophy existed only to perfect human nature and to guide human life, and that in order to secure this end it must elicit and cultivate three virtues or excellences: it must train the understanding to distinguish the true from the false, the useful from the useless, must enable the intellect to penetrate into the nature and trace the order of the universe, and must regulate the will in the practice of what is good; in other words, it must be a Logic, Physics, and Ethics,—a Logic to guide the reason, a Physics to explain the world, and an Ethics to rule the moral life. Each of those disciplines was deemed to include two sciences. Logic was not only the science of correct thinking, but of the correct expression of thought, and so comprehended both Dialectic and Rhetoric; Physics was both a Cosmology and a Theology, Deity being regarded as not separable from the world, but the active and formative power immanent in it; and Ethics embraced Morals and Politics. The Stoics were not agreed as to the order in which Logic, Physics, and Ethics ought to stand. They commonly placed Logic first, but were much divided as to whether Physics should precede or follow Ethics. Logic they likened to the bones and sinews of the animal body and to the shell of an egg, but while some thought Physics was like the flesh of the beast and white of the egg, and Ethics like the soul of the one and the yolk of the other, others represented Ethics as the flesh and white, and Physics as the soul and yolk. The Epicureans accepted the same threefold division of science, but without differing among themselves as to the order of the divisions. They were still more narrowly and exclusively practical than the Stoics; they looked on philosophy merely as the power which conducts men to happiness, and as worth attention only in so far as it contributes to render existence agreeable; hence, Logic they confined to an investigation of the criteria of truth, and cultivated simply as necessary to Physics, and Physics they entirely subordinated to Ethics, valuing it only as the means of delivering the mind from superstitious beliefs which disquiet and embitter the life.

It is unnecessary to criticise this distribution of science either in its Stoic or Epicurean form. It is very obvious that it finds no proper place for, if it does not expressly exclude, metaphysics, mathematics, psychology, and theology; and, in fact, that it excludes at least as much as it includes. It received, however, a wide acceptance, rivalling, and perhaps even exceeding, in its diffusion the Aristotelian classification. It prevailed among the scholastics, and has found favour even with Descartes, Locke, Kant, Herbert, and Hegel, although they have, of course, suggested certain real or supposed improvements. It will, therefore, come before us again in later and more elaborated forms.

Varro. Cicero has no claim to a place in this history, but his contemporary and friend, the learned and indefatigable Varro, is entitled to be mentioned as, in all probability, the first who composed a kind of inventory or encyclopædia of the sciences. Like all but two of the 490 works which he wrote, his treatise *Libri novem disciplinarum* has been lost for ages, but it exerted an influence, through the writings of Capella, Cassiodorus, and others, on many generations to which it was unknown. The nine disciplines of which he treated were the seven so-called "liberal arts," with the addition of medicine and architecture.

In the fifth century of the Christian era, Martianus Capella wrote his bizarre encyclopædic romance, the *Satyricon*. Two books describe the marriage of Mercury and Philology, the daughter of Phronesis, and the remaining seven are devoted to the seven attendants on the bride, the seven liberal arts,—Grammar, Dialectic, Rhetoric, Geometry, Arithmetic, Astronomy, Music. **Capella.**

Somewhat later Cassiodorus treated of the same departments of knowledge in his *De artibus et disciplinis liberalium litterarum*, grouping together Grammar, Dialectic, and Rhetoric, as *Artes* or *Scientiæ Sermocinales*, and Arithmetic, Geometry, Astronomy, and Music as *Disciplinæ* or *Scientiæ Reales*. Capella and Cassiodorus definitively established the educational curriculum for the studious youth of medieval Europe. It has to be remembered, however, that it was only a preparatory course. The studies which it comprised were all regarded as ancillary to a higher science, as so many steps and supports leading up to the knowledge of divine things, the mistress science, Theology. **Cassiodorus.**

They were grouped into what was called the

Trivium and Quadrivium.

Trivium and Quadrivium; the former comprehending Grammar, Dialectic, and Rhetoric, and the latter Arithmetic, Geometry, Astronomy, and Music. The general thought which underlay this division of studies was that those of the lower order were conversant with words, those of the higher with things; that the former were, as they were often termed, *scientiæ sermocinales*, the latter *scientiæ reales;* or, otherwise, that the former were *Logica*, the latter *Mathematica*. The Trivium corresponded likewise to the Logic, and the Quadrivium to the Physics of the Stoics. Ethics was generally included by the Scholastics in Theology, although it was sometimes given a place apart. It was usual for students to pass slowly through the Trivium and rapidly through the Quadrivium, and not uncommon for them to omit the latter altogether, so as to pass at once from logical and verbal studies to what was then the science of most engrossing interest. This, more than any other fact, perhaps, is explanatory of Scholasticism. The scholastics were men whose minds were nurtured on words divorced from things and on the forms without the realities of knowledge. Even the medieval so-called "real sciences" were essentially formal sciences; Arithmetic and Geometry manifestly so, and Astronomy and Music less plainly yet, in the main, indubitably so, as the physical bases and material contents of both these sciences were almost wholly ignored. No wonder, therefore, that so many of the representatives of scholasticism should now seem to us, as we look back upon their exertions, like "metaphysic mills vigorous in grinding the air."

Isidore of Seville (560-636), a celebrated Spanish bishop, and two illustrious Englishmen, the Venerable Bede (673-735) and Alcuin (736-804), greatly contributed to give currency and authority to the scheme of classification of the sciences introduced by Capella and Cassiodorus. Isidore did so by the work entitled *Originum s. Etymologiarum Libri xx*, which at the time of its appearance, and for several centuries afterwards, was supposed to form a complete encyclopædia of all extant departments of knowledge. It was the chief source from which in those times general information was drawn, and had there been no such book, the darkest period of the medieval world would have been even darker than it was. The author's scheme and description of the sciences are contained in his first three books, and the order of their arrangement runs thus: (1) Grammar, (2) Rhetoric, (3) Dialectic, (4) Arithmetic, (5) Geometry, (6) Music, (7) Astronomy, (8) Medicine, (9) Jurisprudence, and (10) Chronology. **Isidore.**

The influence of Bede, owing to his zeal for acquiring and diffusing knowledge, his piety, his authorship of such a work as the *Historia Ecclesi-* **Bede and Alcuin.**

astica Gentis Britonum, and the compends which he wrote to facilitate for students a mastery over various *disciplinæ*, could not fail to be strong and of the same character and tendency as Isidore's. Alcuin, doubtless, owed much to what Bede had been and done, but he was called to work in a far wider sphere. Fortunately, he was well prepared for his mission in life by an admirable and appropriate education in the renowned schools of York, and when he became the friend and preceptor of Charlemagne he zealously sought to have similar schools founded throughout that monarch's wide empire. The king was his first pupil, gave him always his complete confidence, and placed him wherever he could be of most use. During the last years of Alcuin's life he was abbot of the famous monastery of St Martin of Tours, and there, as he had done in other positions, he gave not only lessons on the Bible, but also on ancient languages, grammar, rhetoric, dialectic, and astronomy.

Between the ninth and twelfth century there was little if anything which here concerns us. It must be remembered, however, that from the twelfth century onwards the scholastic doctors, although not independent students of the sciences, or competent to organise satisfactorily the system of the sciences, knew all that Aristotle had taught, much besides which the Jews and Arabs had added, and the vast body of doctrines which had been gradually derived from the statements or suggestions of Scripture. Only minds of the largest capacity could contain those stores of thought and learning possessed by an Albertus Magnus, a Thomas Aquinas, or a Dante.

The most characteristic medieval attempts at classification of the various kinds of knowledge were those which subordinated all secular studies to theology, and represented the former as so many stages by which the soul might gradually raise itself to communion with the Divine. It may suffice to indicate the character of three such attempts, viz., those of Hugo of St Victor, St Bonaventura, and Vincent of Beauvais. Mysticism was a prominent feature of all three, and the mysticism was of a kind which has been appropriately called *Latin*, in order to distinguish it from the earlier *Greek* mysticism of the pseudo-Dionysius and Scotus Erigena and the later *German* mysticism of Eckhart, Tauler, and Thomas à Kempis. In all three stages medieval mysticism was prominent, and naturally so as a much-needed counterpoise to the crude and coarse views, the empiricism, dogmatism, and formalism so prevalent in the medieval world.

Hugo of St Victor.

The classification of Hugo of St Victor (1096-1141) is to be found in his *Eruditio didiscalica.*

Three books in that work treat of worldly sciences, and four of sacred and ecclesiastical history. The former are of most interest. The main object of the entire work, however, is distinctly avowed to be to serve as a propedeutic to theology. All kinds of secular knowledge are held to be of right subordinate and auxiliary to religion. The entire scheme of classification is comprised in three classes or divisions. First, there are the *theoretical sciences.* These include,—*physics* (which is occupied with what is temporal and material),—*mathematics* (which is represented as comprehending the whole four divisions of the *quadrivium*, not merely arithmetic and geometry but also astronomy and music),—and above all *theology* (the object of which is the eternal and divine, and in which alone the reason and heart can find their full satisfaction). Secondly, there is the division of *practical sciences.* It is held to consist of *ethics*, *economics*, and *politics*. And, thirdly, there is a sevenfold distribution of so-called *mechanical or technical arts.* They are arranged in the following order,—*weaving*, *smith-work*, *navigation*, *agriculture*, *hunting*, *medicine*, and the *histrionic art.* When one considers that Hugo was a thorough recluse, of a feeble and sickly constitution, and who is said to have been only once away from his monastery, it must seem marvellous that he should have been able to acquire so much knowledge as he did of such arts as those mentioned.

He did not include psychology in his classification, but he was a student of psychical facts. As he assigned the different faculties of mind to different divisions of the brain, he may be held to have so far anticipated the phrenology of Gall and Spurzheim. In that, however, he was not original. Phrenology should be regarded as not a modern but a medieval invention.

Bonaventura.

The "Seraphic Doctor," St Bonaventura (1221-1274), wrote a treatise entitled *De reductione artium ad theologiam*, in which he sought to refer the varieties of knowledge to the one source of truth—the Father of light. Cognitions he distributed into *artificial*, *natural*, *intellectual*, and *revealed*, according to the character of the Divine illumination in which he supposed them to originate; for, in this view, there are four kinds or degrees of light,—the *external light*, by which we learn the mechanical arts,—the *inferior light*, which shines through the senses, and by which we apprehend individuals or things,—the *internal light*, the reason, which by reflection raises the soul to intellectual things, the universals in conception,—and the *superior light*, the light of grace, which reveals to us sanctifying virtue, and elevates us to universals as they are in their reality—*i.e.*, in God himself. It is, according to Bonaventura, from the internal light that theoretic science or philosophy flows, and such science or philosophy may be threefold, natural, rational, and moral,—the natural including the three sciences of physics, mathematics, and metaphysics,—the rational those of grammar, logic, and rhetoric,—and the moral those of ethics, economics, and politics.

Vincent of Beauvais, R. Bacon, and Dante.

A contemporary of Bonaventura, Vincent of Beauvais, was the author of a very learned work of an encyclopædic nature, the *Bibliotheca mundi*, otherwise known as the *Speculum quadruplex*, since the first part was meant to be a "mirror of nature" (speculum naturale); the second a "mirror of doctrine" or science (speculum doctrinale); the third a "mirror of history" (speculum historiale); and the fourth a "mirror of morals" (speculum morale). In the same century Roger Bacon did noble service to the cause of science by insisting on the regard due to experience, and enlarged men's conceptions of its domain by his advocacy of linguistic, optical, and experimental studies.

Some of our readers will recall to mind how Dante in his *Convito* has represented the distribution of the sciences as corresponding to the divisions of the heavens. Heaven in general is science in general,—science abstract and undivided,—and as there are ten heavens, so are there ten spheres of science. The seven heavens nearest to the earth are those of the planets, and the planets in ascending order are as follows,—the Moon, Mercury, Venus, the Sun, Mars, Jupiter, and Saturn; to them correspond the seven sciences of the Trivium and Quadrivium; and mysterious analogies—so the poet, with an imaginative subtility impossible to describe, seeks to prove—exist between each planet and the science of which it is a symbol,—between the Moon and grammar, Mercury and dialectics, Venus and rhetoric, the Sun and arithmetic, Mars and music, Jupiter and geometry, Saturn and astronomy. Above those planetary heavens are three others, the heaven of the fixed stars, the crystalline heaven, and the heaven of eternal rest, the all-embracing empyrean, not in space but formed solely in the primal Mind; and these heavens represent the highest sciences,—the starry sphere corresponding to physics and metaphysics united, the crystalline to moral philosophy, and the empyrean to theology.

II. FROM THE RENAISSANCE TO KANT.

We must come down to the Renaissance period Poliziano. before we meet with any better schemes of scientific co-ordination. The *Panepistemon* (published in 1491) of the renowned poet and classicist, Angelo

Poliziano, was merely a prelude to more serious attempts. It delineates the tree of knowledge as dividing into the three great branches of *revelation*, of *discovery*, and of *divination*. To revelation corresponds positive theology,—the theology which springs from the fountain of inspiration. To discovery or invention corresponds philosophy, of which the general divisions are these three,—1. *Spectativa*, theoretic or intuitive, including mathematics, physics, psychology, and ontology with natural theology; 2. *Actualis*, practical, comprising ethics, economics, and politics; and 3. *Rationalis*, rational, conversant with grammar the art of expression, history the art of narration, dialectics the art of demonstration, rhetoric the art of persuasion, and poetics the art of intellectual delectation.

Nizolio. There is some originality in the scheme of classification propounded by Mario Nizolio in his *De veris principiis et vera ratione philosophandi contra pseudophilosophos* (1553). Nizolio was a keen opponent of scholasticism, an extreme nominalist, and a decided positivist almost three hundred years before Comte. He held that metaphysics was either false or useless, and to be excluded from among the sciences (*partim falsam, partim inutilem et supervecuam . . . ab omni artium et scientiarum numero removendam*). He equally rejected dialectics and sought to retain only a logic which would concern itself simply with experience, induction, and the simple, clear, and correct use of words. He laid great stress on language, holding thought and speech to be related as soul and body. Hence he represented the tree of the sciences and arts as primarily dividing into the two branches of Philosophy and Oratory, the former tending to wisdom and the latter to its appropriate expression. Philosophy he distributed into *natural* (Physics) and *civil* (Politics),—natural philosophy including geography, meteorology, physiology, and even theology,— and moral philosophy comprising ethics, politics in the special sense of the word, economics, jurisprudence, &c. Under *Oratoria* he ranked all disciplines conversant with speech and composition, *e.g.*, grammar, rhetoric, poetics, and history. At the same time he admitted that numerous departments of knowledge and practice, such as the various branches of mathematics, the mechanical arts, the fine arts, and medicine, could not be included simply and entirely under any one of these three great divisions — Physics, Politics, Oratory — but must be referred to two or even to all of them.

Campanella.

Now we reach Thomas Campanella (1568-1639), who was one of the best representatives in Italy of that great movement of philosophical reform which in the same age produced DesCartes in France and

Bacon in England. Like these two great men he refused to be the slave of the past or to bow down to authority, — summoned the real and reputed doctrines of Aristotle before the bar of reason in order to be tried and tested by their conformity to, or deviation from, nature,—and sought, by substituting experience and induction for dogmatism and *a priori* reasoning, to reconstruct the whole edifice of science,— while, by the courage with which he braved danger and the patience with which he endured persecution, he displayed a strength of soul of which both were destitute and which entitles him to a place in the foremost rank alike of the heroes and martyrs of all time. Campanella, as well as his great English contemporary, endeavoured not only to recall men from an old and false to a new and true method of scientific inquiry, but to map out the provinces of knowledge according to their natural order and relationship. It must be admitted, however, that in this part of his task his services were less brilliant than those of Bacon; that he has not lavished on it the same intellectual wealth; or indicated with the same clearness of vision on his chart of the intellectual world where there are lands to discover; or given utterance to the same magnificent prophecies respecting the future of science. But if his conceptions were not so large and magnificent, neither were they so vague and confused. The principle of his classification was also sounder, inasmuch as he did not set out from a purely subjective position, but aimed at an objective arrangement: in other words, he attempted to classify knowledge not according to the faculties conversant with it, but according to its own nature.

According to Campanella, all knowledge is latent and in germ in sensation,—*sentire est scire*,—but it can only be realised and rendered explicit by intellection ascending from the immediate to the remote, from the known to the unknown, from perception to theory. The foundation, consequently, of all science is history, and as history is either divine or human, the sciences must be divided into divine and human. God is the truth, and all truth must be received from him, but he gives truth in two ways,—he places the book of nature before our eyes, and he speaks to us through the prophets and in our own hearts. Revelation and nature, these are the two sources of all knowledge, the primary divine autographs of which all human systems are but the imperfect and inaccurate copies, and with which they need to be constantly compared to see if they contain anything false. On revelation theology must be built; on nature, micrology. Micrology in its turn is divided in a twofold way, into natural and moral science; the principal branches of the former being geometry, cosmography, astronomy, astrology, and medicine; and of the latter, ethics, politics, and economics, with rhetoric and poetic as auxiliaries. All these sciences, however, treat of particular objects, and there must be another which treats of the universal. They are but parts of a whole; and there must be a study which shows how they are so concentrated and co-ordinated as to form the whole, and what principles pervade and unify them. This study is metaphysics. Its office is to supply principles to all the arts and sciences, and it comprehends a threefold inquiry, namely: (1) into principles of knowledge, (2) into principles of existence, and (3) into principles of action.

Thus Campanella surveyed the domain of science and mapped out its provinces. It is unnecessary to criticise its details, its subordinate divisions, and its delineations of the limits of the special sciences. These, of course, were not, and could not be expected to be, correct. It is of more importance to note that there is hardly a part of the scheme — scarcely a science included in it — on which Campanella has not written with learning and ingenuity; that in holding that a classification of the sciences ought to have regard to their objective aspects, their own natures, their inherent characteristics, he took up the only right position; and that in representing theology as overlying and metaphysics as underlying all the other sciences, and the intervening sciences as composed of two series of sciences, he made a remarkable approximation to a system of co-ordination of the sciences true at least in outline.

DesCartes has not entered on the subject under consideration in any formal or elaborate manner. The most explicit passage regarding it in his writings is the following: "When a man has acquired some skill in discovering truth, he should commence to apply himself in earnest to true philosophy, of which the first part is Metaphysics, containing the principles of knowledge, among which is the explication of the principal attributes of God, of the immortality of the soul, and of all the clear and simple notions that are in us; the second is Physics, in which, after finding the true principles of material things, we examine, in general, how the whole universe has been framed; in the next place, we consider, in particular, the nature of the earth, and of all the bodies that are most generally found upon it, as air, water, fire, the loadstone, and other minerals; in the next place, it is necessary also to examine singly the nature of plants, of animals, and above all of man, in order that we may thereafter be able to discover the other sciences that are useful DesCartes.

to us. Thus, all Philosophy is like a tree, of which Metaphysics is the root, Physics the trunk, and all the other sciences the branches that grow out of this trunk, which are reduced to three principal, namely, Medicine, Mechanics, and Ethics. By the science of Morals I understand the highest and most perfect which, presupposing an entire knowledge of the other sciences, is the last degree of wisdom."[1]

In the context DesCartes informs us that he meant by Philosophy "all that the human mind can know," so that his distribution of Philosophy must be regarded as a distribution of all knowledge. Logic, indeed, he did not include, although he had been speaking of it immediately before, because he looked on logic from an altogether practical point of view, so that it was in his eyes not a part, but the method, of philosophy. Notwithstanding this, his division was nearly the same as that generally adopted by his followers—*e.g.*, by Sylvain Regis, Clauberg, Geulinx — viz., a fourfold division into Logic, Metaphysics, Physics, and Ethics.

Baconian survey.

The Baconian survey of the sciences is a very celebrated one. I venture not to pronounce it unworthy of its fame, although I cannot regard even its leading divisions as accurate. If not a particularly accurate, it was a comprehensive and

[1] Preface to the *Principles of Philosophy*.

attractive, sketch of the intellectual world, indicating in a striking way, difficult to forget, not only what provinces had been acquired by the human mind, but where, and in what manner, new conquests were still to be made. It is difficult to judge what importance Bacon himself attached to it; probably he valued it chiefly because it afforded a convenient framework within which he could arrange his criticisms and counsels regarding each separate science, and his suggestions as to how the "deficiencies" in the literature, learning, and science of his age might be supplied. But whatever was his own estimate of it, Diderot and D'Alembert believed that they could not do better than, in the main, adopt it as the basis of the French Encyclopædia. "If we emerge from this vast operation," wrote the former of these authors in the Prospectus, "we shall owe it mainly to the chancellor Bacon, who sketched the plan of an universal dictionary of sciences and arts at a time when there were not, so to speak, either arts or sciences. This extraordinary genius, when it was impossible to write a history of what men already knew, wrote one of that which they had to learn." A circumstance so remarkable as that the famous French Encyclopædists of the eighteenth century should derive from Bacon's scheme the plan and guiding principles of their gigantic work has naturally of itself drawn much attention to that scheme.

It is a scheme which rests, as I have already observed, on a subjective foundation. Its basis is a division of the faculties of the rational soul. These, according to Bacon, are three,—Memory, Imagination, and Reason. "The sense, which is the door of the intellect, is affected by individual objects only. The images of those individuals—that is, the impressions received by the sense—are fixed in the memory, and pass into it, in the first instance, entire as it were, just as they occur. These the human mind proceeds to review and ruminate on; and, thereupon, either simply rehearses them, or makes fanciful imitations of them, or analyses and classifies them. Therefore from these three fountains—Memory, Imagination, and Reason—flow these three emanations—History, Poesy, and Philosophy; and there can be no others."

Memory, then, which accumulates facts, gives rise to History, which is either Natural or Civil —either of the works of nature or of the works of man. Natural History subdivides into the history of generations, of præter-generations, and of the arts, since nature is, "(1) either free, proceeding in her ordinary course, without molestation; or (2) obstructed by some stubborn and less common matters, and thence put out of her course, as in the production of monsters; or (3) bound and wrought upon by human means, for the production of things artificial." Civil History, in general, subdivides into literary, sacred or ecclesiastical, and civil history strictly so called; the first treating of the progress of literature and learning, the second of the church, prophecy, and providence, and the third of the fortunes of states.

Imagination operates on sensible materials, combining, magnifying, and idealising them at pleasure, and so gives rise to poetry, which, according to Bacon, is simply feigned history, verse being but a character of style. Poetry subdivides into— 1. Narrative Poetry, "a mere imitation of history, such as might pass for real, only that it commonly exaggerates things beyond probability"; 2. Dramatic Poetry, "history made visible, for it represents actions as if they were present, whereas history represents them as past"; and 3. Parabolical Poetry, "typical history, by which ideas that are objects of the intellect are represented in forms that are objects of the sense."

Reason operates on things by analysis and classification, by abstraction and generalisation, and so produces philosophy. But philosophy is not inclusive of all science; it must be distinguished from the knowledge due to revelation—from theology. Theology descends from heaven, philosophy springs from the earth; theology is derived from divine inspiration, philosophy from external sense. At the same time, the knowledge based on revelation may be distributed in the same way as that based on natural perception. "Nor do I think that any other division is wanted for Theology. The information derived from revelation and the information derived from the sense differ, no doubt, both in the matter and in the mode of conveyance; but the human mind is the same, and its repositories and cells the same. It is only as if different liquids were poured through different funnels into one and the same vessel. Theology therefore consists either of Sacred History or of Parables, which are a divine poesy, or of Doctrines and Precepts, which are a perennial philosophy. For as for that part which seems supernumerary, namely, Prophecy, it is but a kind of history: for divine history has this prerogative over human, that the narration may be before the event as well as after."

Division into Theology and Philosophy.

The first division of the sciences, according to Bacon, is into Theology and Philosophy; but in Theology is not included Natural Theology, which is regarded as a part of Philosophy. "Philosophy," he says, "has three objects, viz., God, Nature, and

Man; as also three kinds of rays — for Nature strikes the human intellect with a direct ray, God with a refracted ray, from the inequality of the medium betwixt the Creator and the creatures, and Man, as exhibited to himself, with a reflected ray: so that it is proper to divide Philosophy into the doctrine of the Deity, the Doctrine of Nature, and the doctrine of Man." These, then, are the main branches of philosophy, but the branches must join in a common trunk; the special sciences must diverge out of a general science, consisting of the axioms common to several or to all of the other sciences, and including an inquiry into "transcendentals, or the adventitious conditions of beings." This general science Bacon would name Primary Philosophy. "As the divisions of the sciences are not like different lines that meet in one angle, but rather like the branches of trees that join in one trunk, it is first necessary that we constitute an universal science as a parent to the rest, and as making a part of the common road to the sciences before the ways separate. And this knowledge we call *philosophia prima*, primary or summary philosophy; it has no other for its opposite, and differs from other sciences rather in the limits whereby it is confined than in the subject as treating only the summits of things."

Primary Philosophy.

The doctrine of Deity or Natural Theology Bacon

Subdivisions of Natural Philosophy.

does not subdivide. The doctrine of Nature or Natural Philosophy he first separates into Speculative and Practical; then, subdivides the speculative branch into Physics and Metaphysics—the one the investigation of efficient causes and matter, the other of final causes and form ; and the practical branch into Mechanics, and what he calls Magic, which answers in some measure to Experimental Science. To Natural Philosophy, Speculative and Practical, he adds Mathematics, Pure and Applied, but merely as an appendix, not as an independent science or distinct division of the sciences.

Human and Civil Philosophy.

The doctrine of Man he divides into Human and Civil Philosophy. Human Philosophy he distributes into a doctrine of the body, a doctrine of the soul, and a doctrine of the things common to the body and the soul. The doctrine of the body is to be divided according to the goods of the body, and therefore comprises four sciences—Medicine, which aims at health; Cosmetic, which has regard to beauty ; Athletic, which looks to strength ; and Voluptuary, what Tacitus calls " eruditus luxus," which is conversant with pleasure. The doctrine of the soul comprehends the doctrine of the Substance of the Soul and the doctrine of the Faculties of the Soul, and the latter again includes Logic and Ethic ; the one treating of the understanding and reason, and the other of the will and affections.

Civil Philosophy he divides into the Art of Conversation, the Art of Negotiation, and the Art of State Policy.

Criticism of Baconian Classification.

The Baconian scheme of classification is now before us. We do not overlook its many incidental merits, although we require to confine ourselves to the rapid indication of its chief defects. The main objection to it, as has been often pointed out, is the character of its fundamental principle. The rational soul does not exercise memory, imagination, reason, so much apart, or in as isolated a manner as is assumed, but together, so that all these faculties co-operate in every department of intellectual activity. Take history as the example. Not even in its lowest form is it a mere product of memory; not even in the case of the most stupid historian is it a mere recollection of facts, but a record of facts selected according to certain real or supposed principles of reason. In a higher form, when it aims to reproduce the life of the past, it involves the most difficult and delicate exercise of imagination; and in its highest form, the form of philosophical history, it requires a most comprehensive combination of mental gifts, and one in which mere memory is very subordinate to reason. Further, history and poetry neither admit of entire separation from science nor of distinct co-ordination with it. They are on a different level from science, and may both be covered by science. There is a science of history. Every fact of every kind of history requires to be explained,—that is, to be brought under the domain of science. Historical knowledge is knowledge on the road to scientific knowledge. The perfect history of anything, the complete exhibition of what, how, and why anything is, must be also the science of that thing. In like manner, poetry in all its forms, imagination in all its workings, art in all its varieties and developments, conform to laws and are explicable by reason, and consequently are subjects of science. There is a science, philosophy, or doctrine of the Fine Arts. Æsthetic is the common name for it.

As to the distribution of science, properly so called, there is obviously much that is arbitrary in Bacon's scheme. Theology is separated from Philosophy with a sharpness and absoluteness for which there is no sufficient warrant. Revelation may proceed from divine inspiration, but theological science must be built up on adequately evidenced facts, and by strictly rational processes, even when its facts have their source in revelation and inspiration. The great mass of the facts recorded and of the truths stated in the writings which Christians accept as embodying a revelation, are facts of history and truths accessible to reason; only a very small percentage of them can be exclusively referred to special inspiration. Philosophy cannot admit, consistently with loyalty to reason, that theology is outside of its domain. The separation of natural theology from other theology is the separation of a foundation from the edifice which it supports. Then, the threefold division of philosophy into the doctrine of Deity, of Nature, and of Man is unsatisfactory, requiring, for example, the body of man to have a science to itself widely distinct from the science which studies the bodies of other animals. It implies that the physiology of the human body is more related to psychology than to general physiology. The bringing together of Physics and Metaphysics as both parts of Natural Philosophy is another error which needs no refutation at the present day; the representing of Mathematics as a mere appendix to Natural Philosophy does so still less. The view given of the relation of Logic and Ethics, although at first sight plausible, will be found on examination untenable.

Alsted's *Encyclopædias.*

The state of knowledge in Bacon's age can probably be more fully and distinctly learned from the *Encyclopædias* of John Henry Alsted than from any other works. The first appeared as a quarto volume of upwards of three thousand pages in 1620; and the second, considerably more elaborate, in two large folio volumes in 1630. Alsted was a clear-headed, learned, logical person, skilful in schematising knowledge, indefatigable in composing compends, and his *Encyclopædia* of 1630 was a highly creditable production both in regard to matter and arrangement. By its rigidly methodical character it is no mere dictionary of arts and sciences, but entitled to the name of encyclopædia, as few so-called encyclopædias have been. It consists of thirty-five books. The first four are preliminary, treating of the intellectual habits involved in the acquisition of learning, the characteristics, order, and divisions of the various departments of knowledge, and the ends and methods of study, its aids, hindrances, &c. The six books which follow deal, under the general heading of Philology, with Lexicology, Grammar, Rhetoric, Logic, the Art of Oratory, and Poetic. Philosophy is divided into Theoretical and Practical. Theoretical Philosophy has ten books devoted to it, since it includes ten sciences: Metaphysics, Pneumatics, Physics, Arithmetic, Geometry, Cosmography, Uranometry, Geography, Optics, and Music; Practical Philosophy four books, because it comprehends the four sciences of Ethics, Economics, Politics, and Scholastic. In the three following books the three "Faculties," of Theology, Jurisprudence, and Medicine, are the subjects of dissertation. Theology is distributed into (1) Natural, (2) Catechetic, (3) Didactic, (4) Polemic, (5) Casuistic, (6) Prophetic, and (7) Moral; Jurisprudence into (1) General Civil, (2) Special Civil, and (3) Ecclesiastical: and Medicine in a way requiring more space than we can afford to describe. The three next books give an account of the mechanical arts. The last seven books are miscellaneous and supplementary: præcipuæ farragines disciplinarum: mnemonica, historica, chronologia, architectonica, critica, &c.

From the *Instauratio Magna* of Bacon the great Moravian educational reformer, John Amos Comenius (1592-1671), derived the conviction that universal wisdom—the sum of all science—might be so arranged and presented that it could be acquired without difficulty by any ingenuous and intelligent youth. This belief in the attainability of a Christian pansophy—of an encyclopædic culture which would surely, easily, and solidly lead up, step by step, from the most obvious facts of sense to the secret things of God revealed through Christ—was one of the chief inspiring motives to those labours which have made his name for ever immortal. The aim of his life was to show how his ideal could be realised by means of pansophic schools and pansophic universities. He expounded his conceptions in the *Didactica magna*, *Prodromus pansophiæ*, *Scholæ philosophicæ delineatio*, and other writings

Comenius.

contained in his *Opera didactica omnia*, 4 vols., Amst., 1657. The reader will find an excellent account of what is essential and of abiding interest in these works in the *John Amos Comenius* of Professor Laurie of Edinburgh. Comenius' sketch of a pansophic university is reproduced by Professor Laurie in the following words: "As all knowledge was to lead to God, and to God as revealed through Christ, Comenius spoke of his encyclopædism as a Christian Pansophy, and gave the 'special titles of the seven parts of the temple of Christian Pansophy.' The first was to show the necessity and possibility of the temple and to give its external structure or outline—to be called the *Templi Sapientiæ Propylæum.* The second part was to give the first approach to a knowledge of all knowable things—a general apparatus of wisdom—in which the highest genera and fundamental principles and axioms were to be exhibited, from which, as the primal sources of truth, the streams of all sciences flow and diverge—to be called the *Porta.* The third part (the *primum Atrium*) was to exhaust visible nature. The fourth (the *Atrium medium*) was to treat of man and reason; the fifth part (*Atrium internum*), of man's essential nature—free-will and responsibility, and the repair of man's will in Christ as the beginning of the spiritual life. The sixth part (*Sanctum sanctorum*) was to be theological, and here man was to be admitted to the study and worship of God and his revelation, that thereby he might be led to embrace God as the centre of eternal life. The seventh part (*Fons aquarum viventium*) was to expound the use of true wisdom and its dissemination, so that the whole world might be filled with a knowledge of God" (pp. 72, 73).

Comenius in the last period of his life yielded to the seductions of mysticism. Another religious encyclopædist or pansophist, Erhard Weigel (1625-1699), went much farther in the same direction. He was a proficient in mathematical science and fancied that everything must be explained mathematically. He became a mystic through his excessive trust in the powers of mathematics, and hence while a mystic he was also a precursor of the Wolfian philosophical rationalism. The conception of philosophy as the universal science, and that all philosophy ought accordingly to be treated by the methods of mathematics, is fundamental in his *Idea totius encyclopædiæ*, *Universi corporis pansophici prodromus de gradibus humanæ cognitionis*, *Ethica Euclidea*, and other works. The organisation of knowledge proposed by Comenius was made with a view to the practical requirements of teaching, and that proposed by Weigel was meant to confirm and illustrate a narrow conception of the nature of scientific method. It was not to be expected, therefore, that either

Weigel.

scheme should have much value in the way of indicating the real relationships of the sciences.

Hobbes. The greatest English philosophical contemporary of Bacon, Thomas Hobbes, also attempted a classification of the sciences, and his classification, although it has been little remarked, is, in reality, very remarkable. While Hobbes had not the philosophical breadth or general wealth of mind characteristic of Bacon, he had far more analytic keenness and subtility, far more deductive vigour and self-consistency, and, in a word, decidedly greater specially scientific capacity. In spite of his dogmatic one-sidedness, few English thinkers have surpassed him in energy or range of intellect in the departments in which his strength chiefly lay. His scheme of the distribution and co-ordination of the sciences is exhibited with characteristic conciseness and precision in ch. 9 of *Leviathan* (1651).

Two philosophical theories mould and control it from commencement to close,—sensationalism and nominalism,—of both of which Hobbes was one of the most strenuous and thoroughgoing advocates. Knowledge, he says, is of two kinds,—of facts and of the consequences of one affirmation to another. The knowledge of facts gives rise to history, and history is either natural history or civil history. The knowledge of consequences gives rise to science, which is manifold according to the diversity of matters considered. Its primary division is into Natural Philosophy and Civil Philosophy, according as consequences are from the accidents of bodies natural or of bodies politic.

Natural Philosophy is, in its turn, divided in a twofold manner, according as the consequences of which it consists are drawn from the accidents common to all bodies, which are quantity and motion, or from the qualities of bodies. Consequences from quantity and motion indeterminate constitute Primary Philosophy; from quantity and motion determined by figure, Geometry; from quantity and motion determined by number, Arithmetic; from quantity and motion of bodies in special, if the larger parts of the world, as the earth and stars, Geography and Astronomy; for special kinds of motions and special figures of bodies, Engineering, Architecture, Navigation, &c. Then, going back to physics or consequences from the qualities of bodies natural, these consequences are either from the qualities of bodies transient, such as sometimes appear and sometimes vanish, whence Meteorology; or from the qualities of bodies permanent. Among permanent bodies are the stars, whence Sciography conversant with their light, and Astrology conversant with their influences; the ether, whence a science of atmospheric fluids; terrestial bodies, which are either non-sentient or sentient. Consequences drawn from parts of the earth without sense are Mineralogy and Botany: the one conversant with the qualities of minerals, and the other with the qualities of plants. Consequences from the qualities of animals are either of animals in general or men in special. If of animals in general, Optics is knowledge of consequences from vision; Music of consequences from sound; and some unnamed science or sciences of consequences from the rest of the senses. If of men in special, then, knowledge of consequences from the passions is Ethics; from speech in magnifying, vilifying, &c., Poetry; in persuading, Rhetoric; in reasoning, Logic; in contracting, the Science of Just and Unjust.

Civil Philosophy Hobbes did not subdivide into more special sciences. He supposed it to be largely his own creation, and that its history might be said to have begun with the publication of his *De Cive* (1646).

Thus it was that Hobbes, with clear and systematic genius, mapped out the various provinces of science. The praise of ingenuity and considerable truthfulness cannot reasonably be denied to his arrangement. It shows a deeper and truer insight into the relations of the physical sciences than the chart of Bacon. At the same time, it is not difficult to see defects in it. Some of these, as, for instance, the absence of psychological science, might be supplied without any alteration of the principles on which it proceeds. Others are irremediable, resulting from those principles themselves. Of this character is the exclusion—deliberate exclusion, not simply omission—of theological science. Hobbes maintained there could be no such science, on the ground that there could be no ideas except of the finite and contingent—that body or matter is alone intelligible; that spirit, being beyond the range of experiment and sense, is beyond comprehension, outside of the domain of science. His philosophy was essentially incompatible with a recognition of the existence of theological science.

The strange and arbitrary way in which Hobbes in his classification deals with moral science may also be noted. Ethics is plainly united in the closest manner with Politics, and yet he separates Politics, under the name of Civil Philosophy, from Ethics, by almost as great a distance as his scheme allows. Civil Philosophy stands by itself—isolated, as the counterpart of Natural Philosophy—and Ethics is made a branch, or rather twig, of Natural Philosophy. Nor is this all; but Ethics, as a science conversant about the passions, is separated from the Science of Just and Unjust, and this last, Hobbes, pushing his nominalism to the utmost, represents as a purely verbal science, since, according to him, contracts are the origin or ground of just and unjust.

Locke. The last chapter of John Locke's *Essay concerning Human Understanding* (1690) treats "of the division of the sciences." Locke rightly judged that the consideration of that subject would be a fitting conclusion to such an inquiry into the origin and nature of knowledge as he had instituted. It is only to be regretted that the consideration given was but slight and superficial. The division adopted was threefold —Physica, Practica, Semeiotica—"for a man can employ his thoughts about nothing, but either the contemplation of things themselves for the discovery of truth; or about the things in his own power, which are his own actions, for the attainment of his own ends; or the signs the mind makes use of both in the one and the other, and the right ordering of them for its clearer information." I. Physics, in the wide sense in which the term is used by Locke, is "the knowledge of things as they are in their own proper being, their constitution, properties, and operations"; it has for end bare speculative truth, "and whatsoever can afford the mind of man any such, falls under this branch, whether it be God himself, angels, spirits, bodies, or any of their affections." II. Practics is "the skill of right applying our own powers and actions, for the attainment of things good and useful." Its chief branch is Ethics, "the seeking out those measures and rules of human actions which lead to happiness, and the means to practise them." III. Semeiotics is the doctrine of signs, and includes Logic, or the doctrine of words, "these being the signs which the mind makes use of for the understanding of things, or conveying its knowledge to others."

This division of science is much the same as that employed so long before by the Stoics. It has, however, even as presented by Locke, obvious and serious defects. Thus, for instance, the grouping together of all sciences the objects of which can be said to be "things," as distinct from "actions" and "signs," whatever be the characters otherwise of these objects, and however great may be the differences in the modes and methods in which they must be apprehended and studied, so far from being helpful towards a true correlation of the sciences, is productive of confusion which tends to render their correlation impossible. Further, either of the first two of Locke's groups includes the other two groups. Thus, if Physics comprehend a knowledge of man and of what pertains to man, it must embrace Semeiotics, which is conversant with man's reasoning and speech; and Practics, which is conversant with his activities. So Practics would include all Physics, since whatever knowledge man can attain of Deity, nature, or his own mind, may be turned to use. Strictly speaking, indeed, Practics ought not to be regarded as a kind or branch of science, but as the application of science. The representation of Logic as merely a doctrine of signs may also be set down as erroneous. It implies an extreme of nominalism of which few will be found to approve. Further, as Dugald Stewart observes, "it is difficult to reconcile one's self to an arrangement which, while it classes with Astronomy, with Mechanics, with Optics, and with Hydrostatics, the strikingly contrasted studies of Natural Theology and the Philosophy of the Human Mind, disunites from the two last the far more congenial sciences of Ethic and Logic." In fact, Locke's discussion of the problem—"the division of the sciences"—is so inferior alike to Bacon's and to Hobbes' treatment of it that one can hardly suppose that he had read what they had written regarding it.

Leibniz. Leibniz, in the last chapter of the *Nouveaux Essais*, criticised the classification of Locke, and easily succeeded, of course, in showing it to be radically defective. In particular, he urged with force the objection that each part of the division proposed might absorb the whole. He provided, however, no substitute for Locke's scheme. It does not help us to be told by him that the truths or doctrines of science may be arranged in three ways, viz.: (1) synthetically or theoretically, according to proofs; (2) analytically or practically, according to ends; and (3) lexically, according to letters or terms. What is wanted is an arrangement of the sciences, not of their parts. Only through the right definition and division, constitution and correlation, of the sciences, can their parts, their component truths or doctrines, be scientifically arranged. Besides, the objection which Leibniz urges against Locke's division of sciences applies equally to his own division of methods of arranging truths, if it be presented as the basis of a classification of truths. Any one of these methods is capable of including all truths. Only one of them can be employed at one time, and whichever method be preferred, the classification of truths which is to be in accordance with its principles will have to be made without any help having been afforded by Leibniz.

In fact, Leibniz had no real sense of the importance or clear conception of the nature of the problem before him. Hence his nearest approach to a classification of the sciences is included in a plan for the catalogue of a library,—*Idea Leibnitiana Bibliothecæ ordinandæ contractior.* Now, the classification of the sciences and the classification of books are so far connected that a good classification of the sciences must be of considerable use to one who wishes to classify books, and that a good catalogue raisonnée of books may well afford assistance to one who would classify the sciences; but the two classifications are nevertheless essentially distinct. The classification of the sciences is a fundamental problem of philosophy, the first step toward the correlation of the sciences, and so toward the positive philosophy of the sciences; the classification of books is merely a practical problem of very limited interest, the convenience of bookish people. The classification proposed by Leibniz is one of books, and therefore, like those of Brunet, Girard, Horne, Lubbock, and the general plans of all classed catalogues, necessarily non-philosophical. His classes are,—1. Theology; 2. Jurisprudence; 3. Medicine; 4. Intellectual Philosophy, which is either Theoretical (Logic, Metaphysics, Pneumatics) or Practical (Ethics and Politics); 5. Mathematical Philosophy, which includes not only Pure Mathematics, but Astronomy, Mechanics, and all sciences specially dependent on vigour of imagination; 6. Physical Philosophy, comprehending Physics Proper, Chemistry, Mineralogy, Botany, Zoology, and all sciences which rest on a knowledge of the things of sense; 7. Philology; 8. History; and 9. Miscellanies. According to this arrangement, all knowledge belonging to the three Faculties of Theology, Law, and Medicine is severed and separated from Philosophy or Science, Philology, and History. Thus, to give only a single example, Ecclesiastical History is expressly withdrawn from History in order to be planted in Theology. Of course, this is most arbitrary and unnatural. It would be a mere waste of time, indeed, to discuss at length any scheme of classification in which the subject-matter is divided both according to 'Faculties' and Sciences.

The Italian philosopher, Giambattista Vico (1688-1744), cannot be said to have proposed any new classification of the sciences, and yet ought not to be altogether ignored. In this, as in so many other regions of thought, his power of profound and prophetic vision revealed itself. He was the first to state and expound as a fundamental law of human development the truth which Comte is often credited with having discovered, but which he merely so exhibited as to secure the general recognition of its importance,—the truth that the entire movement of society must correspond to that of knowledge, the preponderant factor of historical evolution being the growth of intelligence. This truth he laid down as the foundation of his New Science not less explicitly or confidently than Comte affirmed it as the basis of his Positive Philosophy. The order of social evolution accord- Vico.

ing to Vico, as according to Comte, is a necessary order determined by the advance of reason. Hence, a law of three periods of history through which all sciences and arts, ideas and institutions, naturally pass. The periods are designated by Vico the Divine, the Heroic, and the Human, and the root of each is described by him as a peculiar mode of conception or form of wisdom. Therefore, he maintains, there are three stages of science, three kinds of nature, three types of character, three epochs of religion, three species of language, of writing, of governments, of natural law, of jurisprudence, &c.

Another equally original idea of his is entitled to be noted here. The "New Science" which he claimed to have founded he maintained to be the central and regulative science. He regarded his discovery of it as not merely an addition to the sciences, but a revolution in the whole system of the sciences, inasmuch as it showed that not metaphysics or physics, but the science of the development of the human mind in history was the fundamental and governing science. In his view the science of history was the most comprehensive science, and all other sciences were rooted or included in it, and had their character and rank determined by their relationship to it. All science, he held, is the production of the human mind; the whole science of any age is only a transient stage in the history of the human mind; the perfect state of a science is but the last period of its history; therefore, the science of history is not merely a special and rather limited science, as we are apt to suppose, but an all-comprehensive science, the true science of the sciences. It is so because the fundamental, constitutive, and regulative principle of all science is not the abstract, transcendent, objective, but the actual, immanent, subjective — the all-productive reason. This was a singularly bold and luminous conception. To demonstrate its truth may be said to have been, consciously or unconsciously, the ultimate aim of all the labours of Kant, Fichte, Schelling, Hegel, and their followers.[1]

With Christian Wolff (1679-1754) and his school, German philosophy passed into a stage of dogmatic rationalism. The general contents of the current philosophy and religion, the teachings of the special sciences, the leading principles and main tenets of Cartesianism, and the distinctive views of Leibniz with certain modifications, were attempted to be systematised and demonstrated by logical deduction of a mathematical rigour and certainty. Wolffianism was essentially encyclopædic. It sought to include and absorb all science. And yet it was Wolff.

[1] See the author's *Vico* in Blackwood's "Philosophical Classics." The book has been translated into Italian.

I

thoroughly one-sided. It ignored the fact that the methods of science must vary with the objects of science; that each science must have its own appropriate modifications of method; that an adequate philosophy can recognise no uniform universal method. It was one-sided also in this respect, that it confounded philosophy with the special sciences. It represented the special sciences as simply sections of philosophy. That is an error so radical as to make unnecessary any other criticism of the Wolffian classification.

Wolff distributes knowledge into historical, mathematical, and philosophical. Philosophy he divides into two great departments corresponding to two fundamental faculties of the soul,—Metaphysics to a *facultas cognoscitiva* and Practical Philosophy to a *facultas appetitiva*. At the same time he treats Logic—chiefly, however, on educational grounds—as antecedent and preparatory to both Metaphysics and Practical Philosophy. In Metaphysics he includes Ontology, Cosmology, Psychology, and Natural Theology. These sciences he regards as following in natural order from more general and simple to more special and complex. In Practical Philosophy he includes Ethics, Economics, and Politics. His follower Baumgarten did good service by vindicating the right of Æsthetics to a place by the side of Ethics.

The Wolffian philosophy was followed by a so-called "Popular Philosophy," which was a continuation of its rationalism, but a reaction from its formalism. The period of the prevalence of this Popular Philosophy was one in which great desire was shown to make the acquisition of science easy. It accordingly abounded in "Introductions," "Outlines," and "Methods." It was a period in which even special Encyclopædias—Encyclopædias of particular departments of knowledge, *e.g.*, Encyclopædias of Theology—began to appear. It was also the period when the want of a propædeutic to the study of the sciences made itself so strongly felt as to give rise to the conception of a special science for its satisfaction and to various attempts to construct such a science; the period in which Gesner, Schade, Mertens, and others sought to raise what they called *Hodegetic or Isagogic* to the rank of a separate and fundamental discipline. It was, above all, the period in which the idea of the organic unity, diversity, and interrelationism of the sciences obtained a universality and clearness of recognition which it had never previously received, although it had at no time since Plato gave it magnificent expression been entirely ignored. It was not, however, a period in which philosophical problems were investigated with depth or thoroughness. As to the problem even of which we are tracing the history it cannot be said to have produced any solution of much value.

III. FROM KANT TO DE TRACY.

Kant. Immanuel Kant (1724-1804), by the publication of his *Critique of Pure Reason*, inaugurated a new and great epoch of philosophy—a philosophy which has had an enormous influence on the higher thought of mankind. He also treated expressly of the very subject we are dealing with, a classification of the sciences, in the chapter of his *Critique* headed "The Architectonic of Pure Reason," and has left elsewhere in his writings various passages supplementary to the views expressed by him in that chapter. He is not therefore to be here ignored. Neither is there, however, any good reason why he should have a large place in any account of a history of our subject.

Science is regarded by Kant as an organism which grows from within, not an aggregate which increases from without. A science, according to Kant, is a system of conceptions unified and distributed by a central and regulative idea; or, in other words, a system organised on what he calls architectonic principles, or constituted by parts which possess an essential affinity and can be deduced from one supreme and internal aim. The idea out of which a science is developed—which is the condition of its possibility, and which determines its form and end—is a constituent element of reason; and hence not only is each science a unity in itself, but all sciences are related as parts of one grand system of knowledge. Knowledge is either rational or empirical. Rational knowledge is based either on conceptions or on the construction of conceptions. In the former case it is philosophy, in the latter mathematics. Philosophy is either a criticism of the powers of reason, Critical Philosophy, or a systematic presentation of the truths given by pure reason, Metaphysic. Metaphysic, again, is either of the speculative or of the practical reason—either a metaphysic of nature or a metaphysic of ethics. The metaphysic of nature divides into two parts—Transcendental Philosophy and Rational Physiology. The former, which may be also called Ontology, presents the system of all the conceptions and principles belonging to the understanding and reason which relate to objects in general, but not to any particular given objects; the latter has nature or the sum of given objects for its subject-matter, and is either immanent or transcendent. Immanent Physiology considers nature as the sum of the objects of experience presented according to *a priori* conditions; and when these objects are those of the external senses it is Rational Physics, when those of internal sense, Rational Psychology. Transcendental Physiology, on the other hand, relates to connections of nature which take us beyond all possible experience, and is, when it embraces nature as a whole, Rational Cosmology, and when it views nature in connection with a Being above nature, Rational Theology. Mathematics, Critical Philosophy, Ontology, Rational Physics, Rational Psychology, Rational Cosmology, Rational Theology, and the Metaphysic of Ethics are consequently the sciences of pure reason. Distinct from, yet related to, Rational Physics and Rational Psychology are to be placed Empirical Physics and Empirical Psychology as parts of Applied Philosophy, the *a priori* principles of which are contained in Pure Philosophy.

This scheme of the sciences suggests various objections. It is not a result of a direct study of the sciences and of their relations to one another, but a consequence of assent to a peculiar metaphysical theory. It is such as was to be expected from treating the problem involved at a wrong place and in a wrong way. The division of knowledge into rational and empirical is radically erroneous, for all knowledge is at once rational and empirical. There is no reason without experience, or experience without reason. That Kant knew this—that he was aware that reason entirely pure, altogether untouched and unaffected by experience, is absolutely ignorant and inactive, and that experience is only constituted by the synthetic activity of reason—far from excusing, is precisely what makes inexcusable his opposing and contrasting, as he does here, reason and experience, rational and empirical knowledge. The division of rational knowledge into Mathematics and Philosophy is as little to be commended. Mathematics is as subject to philosophy, as much comprehended within the sphere of philosophy, as any other science or group of sciences. Philosophy has to deal with the construction of conceptions as well as with conceptions themselves, for it has to treat of the methods of science not less than of its principles. It is universal science. Then, the place which Kant gives to Metaphysic is quite exorbitant and extravagant. In fact, he assigns to it and Mathematics the whole world of science, properly so called. Pure thought—thought which may have a relation to experience, but borrows nothing from it—is represented as able to establish and construct all science worthy of the name, and likewise to lend out of its fulness to empirical studies the principles which alone give them a sort of delusive appearance of science. But neither Kant nor any one else has demonstrated that reason has such a strength and wealth of power, or is more than a faculty or mental instrument of discovering truth about the universe in and through experience.

During the latter half of the eighteenth century

there were, as already indicated, owing to the influence of Wolffianism, Kantianism, and the Aufklärung, a considerable number of attempts made by German authors to give easily intelligible encyclopædic surveys of the whole field of science.

Sulzer. J. G. Sulzer was one of the most highly appreciated German authors of his day. His *Short Summary of all Sciences* mentioned above was not unworthy of his reputation. It was, however, of far less importance than his *General Theory of the Fine Arts* (*Allgemeine Theorie der Schönen Kunste*), first announced in 1760, and published only in 1771-74. Baumgarten, it is true, by the publication of his *Æsthetica* (2 vols., 1750 and 1759), preceded him, and had the honour of first adding Æsthetics to the list of the sciences. A scarcely less honour, however, seems to have been due to Sulzer, as his work apparently was the first in which there was given a comprehensive view of the fine arts (literary included) in their various relationships. For more than half a century he was considered in Germany the chief authority in æsthetics, and that even by those who differed from him in important respects. Sir William Hamilton, to the close of his professional career, and while criticising the psychological basis of Sulzer's views on æsthetics, acknowledged those views to be the best he was acquainted with, his own and Aristotle's excepted.[1] Yet the author of a generally able and admirable *History of Æsthetic*, Professor Bosanquet, has not once mentioned in it even the name of Sulzer. That is surely a strange and large omission. In any general history of æsthetic Sulzer's contribution to the science of æsthetic, instead of being overlooked, ought to have a considerable and prominent place. The general distribution of the sciences proposed by Sulzer was one in which they were referred either to the faculty of knowledge or the faculty of feeling. The inadequacy of it for the purpose intended will now, in all probability, be universally recognised.

Gesner and others.

Gesner's *Primary Lines of Introduction to all Learning*, Meinecke's *Synopsis of all Learning*, Klügel's *Encyclopædic Survey of the different kinds of Knowledge and Science*, Roth's *System of the kinds of Human Knowledge and Science*, and Von Berg's *Essay on the Foundations and all parts of Science*, were all in their day well-appreciated works. Their authors felt themselves to have a mission, endeavoured not unsuccessfully to write with clearness and simplicity, and largely contributed to diffuse throughout Germany desire for a many-sided culture. While aiming, however, at an encyclopædic knowledge they

[1] *Metaphysics*, vol. ii. pp. 467-471.

arrived at no satisfactory classification or co-ordination of the sciences.

Krug. W. T. Krug (1770-1842) was an even more influential author than any of those just mentioned. He had carefully studied the philosophies of both Wolff and Kant without making any surrender of his own independence of mind. And although he did more than any one else to popularise many of the views of the latter, he freely criticised others, and is justly enough classed as only a Semi-Kantian. He was acknowledged to have a very wide acquaintance with almost all the recognised sciences. As regards the fundamental science of Logic, his opinions were both more accurate and more advanced than those of Kant himself, a fact which may so far explain why Sir Wm. Hamilton in his Lectures on Logic made far longer and more numerous quotations from him than from Kant, or indeed from any other logician except Aristotle. Although not so ingenious or profound as Kant, he was very worthy of the successorship to his chair. I have thus far mentioned only his 'Lecture' (Vorlesung) of 1795, as it only had appeared early enough to be in the eighteenth century. A mere 'Lecture,' however, could only be of slight value in comparison with his *Outline of a New Organon of Philosophy*

(1801), his *Handbook of Philosophy* (1803), and his *magnum opus*, his *Universal Handbook of the Philosophical Sciences* (5 vols. 1827). Hence I must in this case cross the boundary between two centuries in order to be able to state what his classification of the sciences was, as it is only on the hither side of the line, only in the later works mentioned, that the scheme was elaborated. I shall do no more, however, than merely state the abstract and general result,—the bare scheme itself. It was as follows :—

The Sciences are either Free or Natural, Bound or Positive, or Mixed.

A. The Free or Natural Sciences are formed solely by the free activity of the human mind, and are reducible to three general groups :—

1. The Empirical, divisible into (*a*) Philological and (*b*) Historical Sciences ;
2. The Rational, comprehending (*a*) Mathematical and (*b*) Philosophical Sciences ; and
3. The Empirico-Rational, which is either (*a*) Anthropological or (*b*) Physical Science.

B. The Bound or Positive Sciences are dependent on authority, and fall into two groups :—

1. The Positive Theological Sciences, and
2. The Positive Juridical Sciences.

C. The Mixed Sciences are those which are theoretically free, but practically and in application subject to and controlled by authority. They comprise—

1. The Politico-Economical, and
2. The Medical Sciences.

Such was the scheme of classification of the sciences ultimately arrived at by Krug. Presented as it necessarily is here, *i.e.*, as a skeleton in all its bareness, it must seem to have little if anything to recommend it. On the contrary, grave objections to it must make themselves felt. One is that no sciences are formed solely by the free activity of the human mind. All of them are to a large extent dependent on the nature of the objects on which free human activity is exercised. Another objection is that no true sciences, either theological or juridical, are *bound* or *positive* in the sense of being dependent on authority. In so far as they are so treated they cannot be truly sciences. The free exercise of rational activity is inseparable from all true science. It holds good of what Krug calls the Mixed Sciences no less than of those which he represents as Bound or Positive Sciences. Only in so far as Political Economy and Medical Studies are free can they be truly sciences, and what is true of them is just as true of Theology and Jurisprudence and all other studies or sciences. In all genuine study science and philosophy, truth and freedom, are inseparable. Notwithstanding these remarks I should greatly regret were any one to infer from them that the works in which Krug's scheme of classification is imbedded are unworthy of study. They are very much the reverse.

In the latter half of the eighteenth century there was nowhere shown so strong a desire for encyclopædic views of the sciences as in Germany. Our English freethinkers of that time showed little interest in the study of scientific or speculative problems. Yet even then England had a Cyclopædia, of a kind now well known and fully appreciated, prior to any other country. I refer to the English Cyclopædia compiled and edited by Ephraim Chambers. It appeared first in 1728, then in 1738, and next in 1739,—the later editions being greatly enlarged by supplementary volumes. It was translated into French and Italian, originated Cyclopædias in all other European countries, and in England became the basis of the greatly extended work of Dr Rees, published in 45 vols. (1802-19). The most widely famed and politically influential of Encylopædias was the French *Encyclopédie, ou Dictionnaire raisonné des sciences, arts et metiers, par une société des gens de lettres* (17 tom. fol., 1751-65). Its two leading contributors were D'Alembert and **Encyclopædic efforts.**

Diderot, both highly gifted men, the former a great proficient in mathematics and physics, and the latter endowed with wonderful readiness of thought and mastery of exposition on any subject. The *Discours préliminaire* was the work of D'Alembert, the article on *Encyclopédie* of Diderot, and the *Prospectus* (afterwards incorporated in the *Discours préliminaire*) of both. In the *Prospectus* the plan of Chambers is admitted to be excellent but the execution is said to be very indifferent. The plan, indeed, alike of Chambers and of D'Alembert and Diderot, was mainly borrowed from Bacon. Necessarily the French *Encyclopédie* with its large *Société des gens de lettres* was much superior in execution to the English *Encyclopædia*, which was almost the work of one man.

D'Alembert.

D'Alembert was unfortunate when he adopted the Baconian scheme of classification as the foundation of his own. The chief alterations made by him on it in his *Preliminary Discourse* have been well indicated by Prof. Fowler in the following passage (*Francis Bacon*, pp. 75, 76): "The places of Imagination and Reason, Poetry and Philosophy, are reversed, so that in the scheme of the Encyclopédie Poetry comes last; the Imagination being regarded by D'Alembert as a more mature faculty (he is, of course, speaking of the creative, not of the merely reproductive Imagination) than the

Reason, and posterior to it in the order of development. Revealed Theology, instead of being treated as co-ordinate with and distinct from Human Learning, is included under that part of Philosophy which is concerned with the knowledge of God, Natural Theology and the Science of evil spirits being the co-ordinate branches. Metaphysics is used in no less than three senses. In one sense, it stands at the head of Philosophy, and has a certain affinity to the Philosophia Prima of Bacon. In another sense, it is employed as the equivalent of Pneumatology, or the science of souls as distinct from bodies, and in this sense is called Particular Metaphysic. Finally, there is a metaphysic of bodies, or general physic, which treats of extent, movement, impenetrability, &c., or the properties common to all bodies. Mathematics is made one of the main divisions of the Philosophy of Nature, instead of a mere appendix, and the mathematical as well as the physical sciences are much more elaborately divided than in Bacon's classification. The various medical sciences, or those which have to do with the care of man's body, are classified on a more scientific basis, and transferred from the Philosophy of Man to the Philosophy of Nature. Morals are divided into general and particular: general ethics being concerned with discussions on the nature of good and evil, on the necessity of being virtuous, &c.: particular ethics with the special duties of the individual when regarded alone, of man in the family, and of man in society, denominated respectively Natural, Economical, and Political Jurisprudence, a similar division being applicable to the conduct of states. Poesy is not confined to Poetry proper, but is made coextensive with the Fine Arts in general."

Those alterations of D'Alembert, however, necessarily failed to improve to any great extent a scheme so radically erroneous as Bacon's,—one of which the root-principle was the separation of three inseparable mental states or faculties. No advantage was gained, or could be gained, by reversing the places of Imagination and Reason, Poetry and Philosophy, as was done by D'Alembert. His putting poetry after history and science, and representing imagination as a more mature faculty than reason, was going farther astray than Bacon had done, and more inconsistent with the testimony of history and psychology. That poetry and art are posterior to history and science is not in accordance with known facts and with the real order of intellectual development. To assign historical studies and their products to memory alone has been already indicated to be erroneous.

The division of history into sacred and secular, ecclesiastical and civil, although a very common one, is also a very misleading one. A history of mankind or the history of a nation is, or at least should be, as sacred and religious as a history of a church or churches. All history is sacred in so far as it is pervaded by the power and spirit of God. No clear, sharp, fixed distinction can be drawn between sacred and secular, ecclesiastical and civil. The larger or so-called civil society is, or at least may be, as well entitled to be deemed sacred and religious as the smaller and so-called ecclesiastical societies within it. The Old Testament is throughout historical, but it certainly never represents history as divisible into sacred and secular, religious and political. There were no ecclesiastical denominations in apostolic times, and the New Testament *ecclesia* never means an ecclesiastical denomination. The *ecclesia* in its distinctively Scriptural sense is not a visible corporation at all, although it manifests itself in all spheres of human activity wherever there is the working of spiritual life. The kingdom of God which is so prominent in the New Testament is certainly not one in which Churchmen are described as having any exclusive or prominent place, but certainly one which from the New Testament point of view is as wide as history itself, because as wide as the whole providential and redemptive work of God as traceable in the history of mankind.

Under the head of 'Memory' D'Alembert adds

to 'Sacred' and 'Civil' History 'Natural History,' and gives a very elaborate distribution of its objects and of the uses to which they might be applied in arts, trades, and manufactures. This was an important addition to the Baconian scheme,—one most creditable to the editor (or editors) of the *Encyclopédie*, and necessarily helpful to those who contributed to it. As 'History' is represented in the scheme to be related only to 'Memory,' so is 'Philosophy' to 'Reason.' Philosophy itself is subdivided into the Science of God, the Science of Man, and the Science of Nature. The first of these is represented as including Natural and Revealed Religion and the Science of Good and Evil Spirits,—a worse than worthless view, such as can hardly be regarded as a serious one.

The Science of Man is divided into Logic and the Doctrine of Morality, and each of these again into Arts. The scheme is elaborate, but to a large extent artificial. The Science of Nature is identified with General Metaphysics, Ontology, or Science of Being in general, and divided into Mathematics and Physics. Mathematics again is divided into Pure and Mixed, and Physics into General, Particular, and Chemistry. Subdivision is carried still farther, and, indeed, too far. On the whole, however, the scheme of classification under the head of Philosophy must, with all its faults, be regarded as a very remarkable and valuable piece of work. It seems impossible to determine how far D'Alembert was aided by Diderot in the elaboration and exposition of it, but there can, I think, be little reasonable doubt that it must have been in the main the work of the former, the more scientific of the two men.

Under the head of 'Imagination' Poetry is divided into Sacred and Profane, then subdivided into Narrative, Dramatic, and Parabolic, and each of these subdivisions into others. But in that there seems to be no merit whatever. In fact, there is not a single science properly so-called included in the section 'Imagination.' The extension given to the term 'Poesie' so as to make it coextensive with 'the Fine Arts in general' was a misapplication of it. Baumgarten had previously found the appropriate term for 'the Fine Arts in general,'—the term *Aesthetik*.

The extraordinary philosophical activity to which Kant's critical investigations into the nature and foundations of knowledge gave rise early in the nineteenth century was applied much more to what was called *the doctrine of science* (*Wissenschaftslehre*) than to direct study of the sciences themselves or of their relations to one another. Fichte, Schelling, Hegel, and their followers felt the necessity of giving to the fundamental problems as to the reality and validity of all that had been assumed to be knowledge better and more constructive solutions than those of Kant. Indeed his solutions seemed to them essentially destructive, and himself 'der zermalmende,'—an even greater 'smasher' of all old theories and doctrines than the Scottish Hume. Hence they felt that their own work must necessarily be not only essentially critical but also essentially constructive, the discovery and proof of a fundamental philosophy or science of knowledge which could not be destroyed like the older theories and systems that Hume and Kant had discredited without finding for them any credible or adequate substitutes. The sciences properly so called could not fail to be influenced by the turn thus taken by speculative thought, nor could they fail to be to a large extent influenced to their disadvantage. Imagination and dreaming got inextricably combined and confused with reason and reality. The minds of the Teutonic philosophers of the time ceased to be conscious of the laws and limitations of human thought. The main result of that was an extraordinary activity in the formation of systems of belief based on some so-called *science of knowledge* (*Wissenschaftslehre*) of a thinker's own invention and maintained by him to be the only true and correct standard of all kinds of knowledge.

Fichte (1762-1814) led the way, and a singularly gifted leader he was. He is certainly entitled to an eminent place in a history of the *Doctrine of Science.* He showed that pure Kantianism—Kantianism as taught by Kant himself—could not be rationally maintained owing to the self-contradictoriness pervading the whole system from beginning to end. With eagle glance he gazed, with eagle swoop he struck, straight at the question around which Kant floundered with whale-like awkwardness, What is the essential unifying factor in all knowledge and in all that is known? He saw that in Kant's teaching there was no such factor, and made it manifest not only that he himself but that Schelling, Hegel, and other eminent thinkers could not consistently rest in a teaching so radically inconsistent as was that of Kant. Hence Fichte must be adjudged entitled to an eminent place in *a history of the doctrine of science* or *philosophy of knowledge.* **Fichte.**

He has, however, no special claim to any such position in *a history of the distribution or classification of the sciences.* He was far from having as wide or accurate an acquaintance with any of the positive sciences as Kant, for example, had; and did little, if anything, in the way of showing how those sciences are related to one another and to the world of science as an intelligible whole. What he really did, or attempted to do, was to represent various kinds of knowledge or action as offshoots of the *Wissenschaftslehre.* It he held to be the fundamental philosophy on which all special studies should be based, to which they must be traced back, and by the spirit of which they should be permeated and vivified; and hence in his various writings he brought into connection with it such subjects as (1) Revelation, (2) Theoretical Philosophy, (3) Practical Philosophy, (4) Law of Nature, (5) Systematic Ethics, (6) Philosophy of History, &c. Indeed, he assumes or affirms all sciences to have their principles in the Science of Knowledge. That, however, does not yield a classification of the sciences.

Schelling (1775-1854) has often been credited with having dealt with the subject under consideration in a rather effective manner, and, in particular, with having anticipated, if not suggested, the solution of it given by Comte. The following words of Morell have been frequently quoted with approval: "The influence of Schelling was not confined to Germany. His attempt to unite the process of the physical sciences in some affiliated line with the study of man, both in his individual constitution and historic development, has also had a very considerable result out of his own country. No one, for example, who compares the philosophic method of Schelling with the 'Philosophie Positive' of Auguste Comte can have the slightest hesitation as to the source from which the latter virtually sprang. The fundamental idea is, indeed, precisely the same as that of Schelling, with this difference only—that the idealistic language of the German speculator is here translated into the more ordinary language of physical science. That Comte borrowed his views from Schelling we can by no means affirm; but that the whole conception of the affiliation of the sciences, in the order of their relative simplicity, and the expansion of the same law of development so as to include the exposition of human nature and the course of social progress, is all to be found there, no one in the smallest degree acquainted with Schelling's writings can seriously doubt."[1]

Since Morell thus wrote documentary evidence has come to light which proves that Comte could not possibly have borrowed from Schelling. It is unnecessary, however, to bring forward that evidence, seeing that the Comtist classification of the sciences has no real connection with the procedure of Schelling affirmed to be, in the main features, identical with it. Schelling's procedure is in no sense a classification of the sciences, and the principle of it is utterly antagonistic to that of Comte. Comte's principle is that of a methodical study of

[1] "Modern German Philosophy"; *Manchester Papers*, 1856.

phenomena. Schelling's is that of the self-movement and potentiation of the Absolute, from the lowest manifestation of so-called matter to the highest activity of reason. The method of Comte is that of science directly studied, generalised, and distributed. The method of Schelling is that of a high-soaring ontology. It is altogether illusory to compare the successive "potences" of Schelling with the fundamental sciences of Comte. Yet it is only just to add that Schelling at all stages and in all phases of his theorising took a keen interest in the sciences, and wrote much of a very suggestive although not infrequently very dubious character. Many a scientist, I imagine, may read, for instance, with considerable pleasure and profit, the *Lectures on the Method of Academic Study*, published in 1803. The subjects treated of in them are the following :—1. *The Absolute Idea of Science;* 2. *The Scientific and Ethical Functions of Universities;* 3. *The Primary Presuppositions of a University Course of Study;* 4. *The Study of the Pure Sciences of Reason, Mathematics, and General Philosophy;* 5. *The Ordinary Objections to the Study of Philosophy;* 6. *On the Special Study of Philosophy;* 7. *Upon* some of the Departments which are to be discriminated from Philosophy—*Specially the Positive Sciences;* 8. *The Historical Construction of Christianity;* 9. *The Study of Theology;* 10. *The Study of History and Jurisprudence;* 11. *Natural History;* 12. *Physics and Chemistry;* 13. *Medicine* (*Pathology*); and 14. *Philosophy of Art* (*Æsthetics*). In his other numerous works he has so far treated of those subjects in relation to absolute science. Although his treatment of them leads neither to a tenable classification nor a satisfactory organisation of the sciences, it has already been, and may perhaps still be, of some value to them.

In the decade from 1806 to 1816 a number of distributions and surveys of the sciences appeared in Germany. It may suffice merely to mention them. Hefter published, in 1806, a *Philosophical Exposition of a System of all Sciences;* Töpfer, in the same year, a *General Encyclopædic Chart of all Sciences,* to which he added, in 1808, a Commentary; Ortloff, in 1807, a *Systematic Distribution of the Sciences,* &c.; Burdach, in 1809, an *Organism of Human Science and Art;* Simon, in 1810, a *Tabular Survey of the Sciences;* the celebrated Lorenz Oken, in 1809-11, a *Handbook of the Philosophy of Nature* (tr. by Tulk for the Ray Society); and Jäsche, in 1816, an *Introduction to an Architectonik of the Sciences.* The works of the first five authors mentioned have quite passed into oblivion. Jäsche is known chiefly as the editor of Kant's *Logic.* Oken is still recognised as a man of genius, but the scheme of science indicated in his *Handbook* has been generally found by those who tried to appreciate it as so original as to be unintelligible.

Hegel. Hegel, in 1817, exhibited, in his *Encyclopædia of the Philosophical Sciences*, a vast system of thought which he believed inclusive of all the fundamental sciences, and necessarily assigning to each of them its appropriate place in the organic and rational whole of knowledge. Judging his work even exclusively from the point of view which here specially concerns us, it must, I think, be pronounced a prodigious advance on those which preceded it, as any one may easily discover for himself by comparing it with the best of the productions mentioned in the previous paragraph. Hegel connects and groups the fundamental sciences in an order which is to a large extent true, and presents a very remarkable exemplification of a most magnificent conception of a Science of the Sciences. He supposes that through the various stages of individual and collective experience and activity described in the *Phænomenology of the Spirit*, and in the "Introduction" to the *Encyclopædia*, consciousness is enabled to rise to absolute cognition, to knowledge of the thought which is all-originative and all-inclusive, to apprehension of the Idea which is the essence alike of nature and of man, the source and explanation alike of existence and of science. The Idea, which is the only appropriate and adequate object of absolute cognition, Hegel believes himself to have attained, and his whole philosophy, as exhibited in the *Encyclopædia*, is an attempt to trace the chief phases and forms of its development. In direction the development is from abstract to concrete, from simple to complex, from barest poverty to fullest wealth of content; in character it is rhythmic, reasoned, dialectic; and the character of the movement determines its direction, its whole course, and ultimate goal, seeing that in affirming itself the thought with which philosophy is conversant likewise denies itself, yet so as thereby, instead of destroying itself, to reconcile itself to itself, and this through innumerable forms which become ever more concrete and comprehensive, until the whole content of the Absolute Idea is evolved. Owing to the very nature of the Hegelian dialectic, the Hegelian philosophy is threefold alike as a whole and in its parts. It must treat of the Idea in itself, in which case it is Logic; or of the Idea in its other or external form, and then it is the Philosophy of Nature; or of the Idea in its return to itself, when it is the Philosophy of Spirit. In like manner the threefold rhythm of the dialectic process causes Logic to resolve itself into the Science of Being, the

Science of the Essence, and the Science of the Notion; the Philosophy of Nature into Mechanics, Physics, and Organics; and the Philosophy of Spirit into the Doctrine of Subjective Spirit, the Doctrine of Objective Spirit, and the Doctrine of Absolute Spirit, the first of which comprehends Anthropology, Phenomenology, and Psychology, while the second deals with Legal Right, Morality, and Ethical Obedience, and the third embraces the spheres of Art, Religion, and Absolute Philosophy. Thus the fundamental sciences are represented as having each a fixed and appropriate place, as bound together by ties of rational affinity, and as the necessary and constituent members of a vast harmonious and organic system of knowledge. Hegel must, consequently, be credited with having made an enormous advance on all schemes of classification of the sciences by mere logical division, external arrangement, or figurate representation. He has aimed at a real incorporation of the special sciences into a general science, at a thorough reduction of them under a comprehensive doctrine, at a correlation of them based on consideration of the entire contents of each. This may well render us averse to dwell on errors of detail in his views. These are neither few nor difficult to discover, and have been often indicated. Hegel has, perhaps, oftener failed than succeeded in defining the limits of the particular sciences. It is only in a very general way that his scheme of co-ordination can be defended. The defects of his Philosophy of Nature are notorious, and the great merits of his Philosophy of Spirit are blended with serious faults. But to ignore the truth and grandeur of his general theory of the correlation and combination of the sciences in critically gazing at such imperfections must be pronounced almost as irrational and unjust as to doubt or deny the brightness of the sun because a telescopic examination shows it to be mottled over with a number of dark spots. Whatever be the faults of Hegel's *Encyclopædia*—although they be even "thick as dust in vacant chambers"—this glory, I think, cannot fairly be denied to it, that there, for the first time, appeared a system of such a character and scope, so vast in its range of conception, so rich in suggestion and doctrine, and so skilfully constructed, as to present to the mind something like what a Science of the Sciences ought to be.

I refrain not only from urging particular objections to the Hegelian scheme of scientific co-ordination, but also those general objections which might be drawn from the nature of the Hegelian Idea, and of the Hegelian dialectic. These objections may be both relevant and conclusive, but they obviously raise the whole question of the truth or falsity of the Hegelian philosophy, which is a question far too large to be here discussed. The late Prof. Harms, in his *Geschichte der Psychologie* (pp. 42-48), has rejected the Hegelian classification especially on the ground that the dialectic process is a form of evolution inconsistent with either the sciences or their objects differing otherwise than in degree, although the facts of experience show that they differ essentially and specifically. It is an objection to which I cannot attribute much weight. It may be difficult to conceive that any process of evolution can produce certain differences, but it is also difficult to show that they may not, and off-hand appeals to experience on the question are to be deprecated. Then, of all forms of evolution, the Hegelian seems to be the one against which the objection must strike with the least force, seeing that the Hegelian dialectic, while a process which goes on without interruption or cessation, is also one of which each stage has a certain essence and peculiar character of its own, each of the three moments or acts included in it being relatively distinct. The evolutionism of Hegel does not attempt, like that of Darwin, and at least like that of contemporary materialism, to explain development entirely by gradation. It affirms unbroken continuity of movement, but at the same time maintains that the movement throughout includes distinctions of nature, not merely differences of degree. The only objection on which I deem it necessary to insist is that a doctrine of the sciences ought to be based on, and built up by, a direct study of the sciences, instead of being drawn out of the bosom of a metaphysical philosophy. It must be reached through induction, not from deduction; through analysis and generalisation, not by synthesis and specialisation; by an upward, not a downward movement. It should be the product of philosophic thought, but of such thought in its first stage of advance on the thought which has produced the various sciences. It is one of the means with which the intellect must provide itself in order to apprehend ultimate and absolute truth. The view that a doctrine of the sciences must be derived from a doctrine of science, and even from a doctrine of Being, is very plausible, yet very erroneous. A doctrine of the sciences undoubtedly implies a doctrine of science, and even a doctrine of Being; but for this very reason it must precede them, and they can only be attained through it. What is first in the order of nature is last in the order of knowledge. To reach the centre of truth, every point which lies between it and the circumference must be passed through. Hegel disregarded all considerations of this kind. He started from what he believed to be truth higher than the truths of science without having made use of the sciences to reach it. He began with philosophy at its highest, sought to work it all out by a uniform method from an absolute first point, and so to incorporate into it the sciences, to assign to each of them its place, and to exhibit their relationships. This I hold to have been a radically erroneous procedure. I must be content, however, simply to state the conviction, having indicated at the commencement of my previous paper what I deem to be the true position and function of a doctrine of the sciences in the organism of philosophy.

De Tracy. Two years before the publication of Hegel's *Encyclopædia* a celebrated French philosopher, A. L. C. Destutt de Tracy, completed a *Cours d'Idéologie* (1801-15, 5 vols.), in which he attempted to trace a plan of the whole edifice of science in accordance with the general philosophical principles of Locke and Condillac. He maintains that the foundation of all science must be acquaintance with the principles implied in the formation of science—the knowledge of how knowledge, which consists of ideas, is obtained from sensations or feelings. Ideology must be, consequently, the fundamental science, and it includes three departmental sciences—Ideology in the narrower sense,

Grammar, and Logic—which treat respectively of the formation, the expression, and the combination of ideas. Then, our means of knowledge may be applied either to the study of what is within or of what is beyond our power—either to the study of the operations of the will or of the properties of nature—and hence there are other two groups of sciences. The sciences which refer to the will are Political Economy, Morals, and Jurisprudence; those which refer to external nature are Physics, Geometry, and Arithmetic.

Such is De Tracy's scheme of classification. Obviously the enumeration of sciences in the second and third divisions is very incomplete, and the arrangement of them careless. The omission of Æsthetics, the Science of History, and especially Theology, cannot fail to be remarked. And even the leading conception of his scheme—the view that the primary science must be a science of the conditions and processes implied in the formation of science—is extremely questionable. How are we to ascertain the conditions and processes of science except through a study of the sciences, and how shall we study them unless they exist? An Ideology not drawn from ideas, a Grammar not dependent on languages, a Logic which does not presuppose the reasonings and methods of science, must be most unworthy to be called sciences.

IV. FROM BENTHAM TO GIOBERTI.

Bentham. Jeremy Bentham and Samuel Taylor Coleridge were the Englishmen who at the period now reached discussed the problem of the classification of the sciences. The former in the fifth appendix to his *Chrestomathia*, first published in 1816, did so. His scheme assumes that "directly or indirectly, *wellbeing*, in some shape or other, or in several shapes, or all shapes taken together, is the subject of every thought, and object of every action, on the part of every known *Being*, who is, at the same time, a sensitive and thinking Being"; that "art and science so run along everywhere together that every division performed on the one may, on any occasion, be considered as applying to the other"; that all the arts and sciences meet in, and proceed from, a central, common, and comprehensive art and science—Eudæmonics; and that the distribution of this art and science into the various arts and sciences ought to be exhaustive, and may be made so through lengthened dichotomous division, continued bifurcate ramification. These assumptions are not to be admitted. The first is the basis of utilitarianism, but denied by all who reject utilitarianism; the second ignores the fact that the points of view of science and of art are so different that every science is of use in several arts, and that every art presupposes several sciences; the third falls with the two assumptions which precede it and on which it rests; and the fourth has been so discredited in every department of inductive study that the belief in the applicability of dichotomous division either to the realities of nature or to their reflections in thought is now justly deemed by scientific men a superstition.

The all-comprehensive art and science of Eudæmonics may be regarded, according to Bentham, specially either as art or science, and the name Eudæmonics may be specially appropriated to the former, and Ontology to the latter. "In every part of the common field, concomitant and correspondent to *Eudæmonics*, considered as an art, runs *Ontology*, considered as a science." Ontology is, therefore, the trunk of the tree of science, while the other sciences are branches of that tree formed by successive bifurcations. The tree itself is, consequently, a Ramean tree. Thus Ontology is divided into Cœnoscopic (Metaphysics) and Idioscopic; Idioscopic Ontology into Somatology and Pneumatology; Somatology into Posology and Poiology, and Pneumatology into Nooscopic and Pathoscopic; and so on, until the result is reached that Poso-scopic Somatics includes Geometry, Arithmetic, and

Algebra; Poioscopic Somatics, Astronomy, Botany, Zoology, Experimental Philosophy, and Technology; Nooscopic Pneumatics, Logic, Grammar, and Rhetoric; and Pathoscopic Pneumatics, Æsthetics, Exegetic Ethics, Private Ethics, and the Political Sciences. The process by which this result is attained is not only long and wearisome, but at almost every stage very questionable. Theology is entirely ignored. Bentham, like Hobbes, supposed it not entitled to any place among the sciences. His whole scheme, indeed, reminds us of that of Hobbes. It is as self-consistent and even more elaborated, but shows less vigour and perspicacity, and more narrowness and pedantry of mind. Its nomenclature is hideous, but ingenious and significant. In the encyclopædic language of Bentham, Arithmetic is Gnostosymbolic, Alegomorphic, Pososcopic, Somatic, Cœnoscopic Ontology; Zoology is Embioscopic, Epigeioscopic, Physiurgic, Pososcopic, Somatic, Idioscopic Ontology; and Rhetoric is Pathocinetic, Cœnonesioscopic, Nooscopic, Pneumatic Ontology. These are wonderful and fearful propositions at first sight or first hearing, but any reader possessed of a little Greek may easily translate them into English, and will learn something by doing so.[1]

[1] Bentham's *Chrestomathia* is contained in vol. viii. of Bowring's edition of his works.

Coleridge divided the sciences into Pure Sciences, which are built on the relations of ideas to each other, and Mixed and Applied Sciences, which are built on the relations of ideas to the external world. The Pure Sciences he subdivided into Formal and Real, the former exhibiting the forms of thought, and the latter treating of Being itself, of the true nature and existence of the external universe, of the guiding principles within us, and of the Great Cause of all. Grammar, Logic, and Mathematics he classed as the Formal Sciences; Metaphysics, Morals, and Theology as the Real Sciences; Mechanics, Hydrostatics, Pneumatics, Optics, and Astronomy as the Mixed Sciences; and the various branches of Experimental Philosophy, the theories of the Fine Arts and of the Useful Arts, and Natural History, with its applications to Medicine and Surgery, as the Applied Sciences. **Coleridge.**

It is not difficult to discover grave defects in this classification. The Real Sciences cannot be Pure Sciences if Coleridge's own definitions of Real Sciences and Pure Sciences be correct. The Mixed and Applied Sciences, if only mixed and applied, have no right to be classed as co-ordinate with the Pure Sciences; and if in any degree distinct and independent sciences, they must be to the same extent either Formal or Real Sciences. Most of them are obviously entitled to be ranked among the so-called Real Sciences. Within the several groups the order in which the particular sciences follow one another is not the most natural order. For all defects of this kind Coleridge himself may not be responsible, as he complained that under editorial revision his work was (to use his own words) "so bedeviled that I am ashamed to own it."[1]

Jannelli. Cataldo Jannelli, a clear-headed Italian author, while endeavouring to correct and develop in his *Cenni sulla natura e necessità delle cosse e delle storie umane* (1817) the doctrine of Vico, dealt, although only to a slight extent, with the problem of the classification of the sciences. While recognising the value of the work achieved by Vico in the *Principii di Scienza Nuova*, and his right to be regarded as the founder of the philosophy of history and the improver of all sciences dependent on that philosophy, he was sufficiently independent to criticise even the central doctrine and most comprehensive generalisation in the great Neapolitan's system of thought. For Vico's *divine, heroic, and human ages* he substituted three ages partly correspondent to and partly corrective of them—namely,

[1] As to Coleridge's classification see the third section of his *Treatise on Method*, prefatory to the *Encyclopædia Metropolitana*, which began to appear in 1817.

the ages of sense, imagination, and reason—and thus with a considerable measure of success improved on his predecessor's description of the stages of human development and distribution of the kinds of knowledge, without rejection of the great idea of a natural law of the development of life in the history alike of the human race and the human individual. His criticism left psychology, sociology, and evolutionism none the less indebted to Vico, while it led him to recognise as of supreme importance the claims of two other sciences — namely, *teleologia* (the science of final causes) and *ideologia* (the science of first causes). His general distribution of the sciences is into *intuitive* or *theoretical* and *operative* or *practical sciences*,— a much too simple classification.

G. D. Romagnosi (1761-1835), a very eminent Italian jurist and publicist, and a wise and independent citizen in a very difficult and critical period of his country's history, gave expression in one of his many writings to what may well be regarded as an extravagant view of the importance of an encyclopædic distribution of the sciences. I quote his words below.[1] So far as I am aware, he

Romagnosi.

[1] *Vedute fondamentale sull' Arte Logica*, Lib. i. Sez. i., § 18: "Un albero enciclopedico delle scienze ben fatto forma l' ultima e la piu grande espressione del logico magistero."

himself made no attempt to supply the sort of classification he so highly appreciated.

Longo. It seems certain, however, from the Appendix to the second volume of Prof. Vincenzo di Giovanni's *Storia della Filosofia in Sicilia*,—a very interesting and every way admirable work,—that the problem must have been long and earnestly dealt with by a Sicilian scientist, the Cav. Agatino Longo. Greatly to my regret I have not been able to obtain his writings on the subject—one which must have occupied his mind more or less for over thirty years. Unfortunately for my purpose Di Giovanni has given no information as to their contents beyond what is implied in their titles, and as they have all been published in Sicily, and for the most part in Sicilian periodicals, generally short-lived and of very limited circulation, I have not been able to obtain them, and must therefore content myself with reproducing a few of the titles given by Giovanni in the work already mentioned, viz. :—

Longo, Cav. A.—

Prolusioni accademiche, lette nell' università di Catania. (La prima di esse ivi stampata nei 1820 presenta una nuova classificazione delle scienze: la seconda inserita nel t. xiii. del. Gior. di scienze lettre ed arti offre partizioni della erudizione e delle arti.

Sul bisogno d' una nuova classificazione delle cognizioni, Cat. 1827, e nel tomo xxii. del Giorn. di scienze lettre ed arti per la Sicilia.

Atlante universalle delle cognizioni, o Tavole sinottiche contenenti la classificazione sistematica delle scienze secondo il metodo naturale. (Di questa grand opera ne dà l' annunzio nel t. xxxiv. di detto Giornale, e nel xiii. dell' Efemeridi sicole.)

Osservazioni sulla Geneografia dello scibile del sign. Pamphilis: nel t. xxxv.

Sul valore del vocabolo Filosofia, ed enumerazione delle scienze che vi s' includono. Cat. 1850.

Delle partizioni della filosofia generale, e dei metodi di classificazione. Cat. 1850.

Father Giovachino Ventura set forth his views on the classification of the sciences in his *De Methodo Philosophandi*, published at Rome in 1828. But his traditionalism, his subjection of reason to authority and of science to faith, his want of secular knowledge and exclusively theological habits of thought, rendered it impossible for him to discuss the theme with much success. He assigned to the encyclopædic tree of knowledge three branches—one bearing *the sciences of authority*, another *the sciences of ratiocination*, and the third *the sciences of observation*. These he represented as coincident with the *Ethics*, *Logic*, and *Physics* of ancient philosophy. That view, it need scarcely be said, is utterly erroneous. It is, however, not more so than **Ventura.**

this other, closely connected with it, that the object of Ethics, thus understood, is God; of Logic, Man; and of Physics, Body. Ethics is divided into Metaphysics and Jurisprudence; Logic into Ideology and Logic strictly so called; and Physics into General and Special Physics. The process of subdivision is pushed to a great length.[1] The self-confidence of the renowned Theatine orator was undimmed by any suspicion of ignorance, and so he mapped out the universe of knowledge with magisterial minuteness. It would serve no good purpose to follow him in details, which are without interest in themselves, and which belong to a scheme false in its principles and misleading in its main lines.

Ferrarese. L. Ferrarese published, in 1828, a *Saggio di una nuova classificazione delle scienze.* It contains good remarks on the importance of a right distribution of the system of knowledge, but the classification which it sets forth is not based on sound principles, and by no means satisfies the necessary requirements. According to Ferrarese, the Science of Man must be the foundation of all the sciences; but he has forgotten to attempt to prove that there can be a Science of Man without a foundation supplied by other and simpler sciences. He classifies the sciences exclusively according to the modes of their helpful-

[1] *Op. cit.*, art. vii., pp. 241-300.

ness to man, on the ground that they will not otherwise be so useful to him as they might be; but in so doing he overlooks that even if this allegation were correct—which it is not—it would be altogether insufficient to establish that the sciences should have their place and rank determined, not by intrinsic, but by extrinsic considerations; not by the nature of the truths of which they consist, but by the uses to which they may be put. To the tree of science he assigns three great branches, because the sciences contribute, in his opinion, to one or other of three great ends — the maintenance of man in health or soundness (*salute*) of body or mind, the furtherance of his perfection, and the prevention of his degradation —although these ends are obviously so closely associated that any one of them can only be realised in the measure that the others are promoted, and that, consequently, to distribute the sciences by them into distinct groups must be futile. The fundamental science Ferrarese calls Anthropography, and he describes it as dividing into Descriptive and Comparative Anthropography. To these two branches the mathematical, physical, natural, and medical sciences are represented as belonging. The third branch begins with Telestics, the general science of the perfection of man, alike as regards his bodily, intellectual, and moral faculties. It is supposed to support Pædagogy, the philosophical, juridical, and moral sciences, history, and literature.

De Pamphilis.

The treatise of Giacinto de Pamphilis—*Geneografia dello Scibile*—was published a year later than that of Ferrarese,[1] and is even more ingenious. It places the centre of the sciences not in man alone, as the work of Ferrarese does, but both in nature and man, since the former is the objective cause, and the latter the subjective cause, and these causes act incessantly on each other. It refers the origin of knowledge to "the reciprocal circular influence" between nature and man, and makes this fact the principle of the division of the sciences. Hence it distributes the sciences into three orders: 1. Objective Sciences, those of the Not-Me; 2. Subjective Sciences, those of the Me; and, 3. Objective-subjective and Subjective-objective Sciences, those of the Me in relation to the Not-Me, and of the Not-Me in relation to the Me; or, in other words, into Physical, Metaphysical, and Moral Sciences. These orders are brought into connection in a somewhat intricate and arbitrary manner, so as to yield such groups as Grammar, Logic, and Morals; Cosmology, Psychology, and Theology; Philoagathy, Philocaly, and Philosophy; Metaphysics, Ontology, and Ideology; all the members of which deal directly with

[1] A second edition appeared in 1869.

the phenomenal, yet imply the transcendental and absolute. It would occupy too much of the space at my disposal to explain and criticise the processes by which these groups are formed. Any reader whose curiosity regarding the scheme is unsatisfied may be referred not only to the work in which it was propounded, but also to the examination of it by Prof. Longo, *Osservazioni sulla Geneografia dello scibile del sig. Pamphilis.*[1]

Neil Arnott.

Dr Neil Arnott, in the introduction to his *Elements of Physics,*—a popular work, of which the first edition appeared in 1828,—divided the whole sum of man's knowledge of nature into Natural History and Science or Philosophy. The former treats of the materials of the universe—*e.g.*, minerals, vegetables, animals; or, in other words, describes the kingdoms of nature. The latter treats of the manners or kinds of motion or change; or, in other words, exhibits the general truths or laws of nature. It ought, in Dr Arnott's opinion, to be distributed into four distinct sciences—Physics, Chemistry, the Science of Life, and the Science of Mind—because all phenomena are referable to four distinct classes,—the *physical*, *chemical*, *vital*, and *mental.* These four sciences "may be said to form the pyramid of Science, of which Physics is the base, while the

1 *Giorn. di scienze lettere ed arti per la Sicilia*, t. xxxv.

others constitute succeeding layers in the order mentioned, the whole having certain mutual relations and dependencies well-figured by the parts of a pyramid." Mathematics "may be considered as a subsidiary department of human science, created by the mind itself, to facilitate the study of the others." Theology is included in the Science of Mind. It was thus that Arnott enunciated an idea of a hierarchy of fundamental sciences closely resembling that of Comte, and, indeed, superior to it in the two points in which it differs from it—namely, in neither representing Astronomy as a fundamental science, nor the Science of Mind as merely a department of the Science of Life. There is no evidence, so far as I am aware, that his anticipation of Comte was due to any acquaintance with the writings of Saint - Simon. He enunciated, however, Saint - Simon's general idea, although only in a very general way. He made no attempt to build on it, as Comte did, a universal philosophy, a science of the sciences. How incompetent he was to perform such an achievement, had he been ambitious enough to undertake it, we may judge from the feeble book he published in 1861, entitled *A Survey of Human Progress.* Yet in this work he developed in some degree the conception just indicated as contained in his earlier one. He, as Dr Bain says, "brought out more decisively the distinction between Sciences and Arts, and between the Concrete and the Abstract Departments of Science." Still distributing that knowledge of phenomena to which he restricted the term Science or Philosophy into the four fundamental Sciences of Physics, Chemistry, Physiology, and Psychology, he represented the knowledge of things or objects called Natural History and the devices or practical applications of knowledge called Art as similarly divisible, so that the departments of Natural History, of Science, and of Art form three parallel and co-ordinate series, Astronomy and Geography corresponding to Physics, Mineralogy and Geology to Chemistry, Botany and Zoology to Physiology, and the History of Man to Psychology, while the Arts must be classified as Mechanical, Chemical, Physiological, and Mental.

We now reach Auguste Comte, than whom, perhaps, no one has done more for philosophy as positive. He owes the high place he holds among philosophers to the power and skill and general truthfulness of his elaboration of the doctrine of the so-called positive sciences as a whole, not to the merits of his treatment of the particular problem of the classification of the sciences. He claimed, but had no right whatever to claim, that he originated the classification which he adopted. If

Comte.

that classification possess any merits, they must be ascribed to Dr Burdin, who conceived it, and to Saint-Simon, who first received and published it; not to Comte, although he showed how much could be made of it. As it is with Comte's name, however, that the classification is almost universally associated, it is in connection with him, and as expounded in his *Cours de Philosophie Positive* (1830-42), that I shall briefly consider and criticise it.

The classification cannot be dissevered from the celebrated so-called "law of the three states." That alleged law, as it is understood and expounded by Comte, means that the human mind in every department of thought and inquiry reaches such relative truth as it can attain, and so enters into the state called positive, or, in other words, arrives at science only by passing through a theological and metaphysical state, both essentially false and conjectural, although both containing some measure of truth and pervaded by a certain nisus toward the certainty of science. Thus apprehended, the law necessarily implies that there can be no true theology or true metaphysics, and that whatever professes to be theological or metaphysical science must be discarded as pretentious delusion. Comte cannot be charged in this respect with want of consistency; he refused to assign either to theology or metaphysics any place among the sciences. That in doing so he most seriously erred I shall not here endeavour to show, as I have elsewhere examined his views on this point at considerable length.[1]

Comte further defined and limited the field of his investigation by excluding from consideration merely composite and derivative sciences. He distributed the sciences into two classes—Abstract Sciences conversant with general laws, and Concrete Sciences conversant with the explanation of particular existing things by means of general laws; and held the former only to be fundamental, and alone to require from the philosopher classification. He thus greatly simplified his task. There can be no doubt that the distinction on which he rested the simplification is a very valuable one. It is now almost universally accepted.

The Abstract Sciences, Comte held, must fall into a single linear series, each member of which has its place determined by its relative simplicity, generality, and independence. This does not prevent them from being divisible into Mathematical and Physical, or the Physical Sciences from being divisible into Inorganic Physics (comprehending Astronomy and Physics proper) and Organic Physics

[1] *Philosophy of History*, pp. 267-278. In ch. x. of my *Historical Philosophy in France* I have treated somewhat fully of the naturalism and positivism in the doctrine of Comte, but not at all of his attempted classification of the sciences.

(containing Biology and Sociology); but it implies that Mathematical Science must precede Physical Science, and that the five fundamentally distinct Physical Sciences must have been evolved in the following order: Astronomy, Physics, Chemistry, Biology, and Sociology. A relatively simple, abstract, and independent science must always precede one which is more special, complex, and dependent.

These views of Comte raise various questions. One is this: Is there, even of sciences of the kind which he calls positive, only one series? Is there not, for example, a Psychical as well as Physical series of such sciences? The material and the mental spheres of existence are conspicuously different and appear to be essentially distinct. The facts on which the physical sciences are built are all observed externally by the senses, while those on which mental science is built must be apprehended by internal consciousness; we cannot observe a single fact of physical nature by introspection, nor a single fact of mind by perception. From this it seems to follow that, although Psychology may possibly be the root of a series of sciences parallel to the Physical Sciences, neither itself nor any science springing from it—as, for example, Sociology—can be included in a series of Physical Sciences. And certainly Comte has not shown this conclusion to be unwarranted. The views he maintained as to the position, character, and method of the Science of Mind cannot commend themselves to any competent student. The arguments from which he inferred that Psychology is merely a department of Physiology, and may even be identified with Phrenology, are singularly weak and irrelevant, and have often been adequately exposed.

There is a still more penetrating question: Is there a fixed line or series either of the physical or psychical sciences? Is there in any group of the sciences a straight line of succession and necessary order of filiation? Comte maintained that there is, while Herbert Spencer, in an essay on "The Genesis of Science," has argued, with great ingenuity and vigour, that there is not; that "the conception of a *serial* arrangement of the sciences is a vicious one"; that "there is no 'one *rational* order among a host of possible systems'"; that "there is no 'true *filiation* of the sciences.'" That Comte's doctrine is very inadequate and inaccurate Spencer seems to me to have conclusively shown. Indeed, a very general inspection of the procedure of the mind in the formation of the sciences must suffice to convince us that Comte has erred in his views as to the filiation of the sciences. The nature of the connection, or so-called filiation of the sciences, must depend on the nature of their formation or genesis. The former must be simple or complex according as the latter is simple or complex. Now Comte supposed the latter to be simple, while in the positive sciences, both physical and psychical, it is really and obviously complex. It is not a single, but a twofold process. In the formation of any of the positive sciences, since a positive science is the explanation of facts by laws, the mind for some time predominantly and always to some extent follows an ascending direction, rising from facts to laws, from sense to science. On this path its instruments are induction and its auxiliary processes, and with their aid it evolves laws of ever-increasing comprehensiveness and simplicity. But the reverse method, the descending order, must likewise be followed. The results of induction become the premisses of deduction. The laws inductively reached yield deductive solutions of problems previously inexplicable. But since the progress of science thus depends not on one process of discovery, but on two processes, the one the inverse of the other, the order of the evolution of the sciences must manifestly be very different from what it would be if determined by a single process, whether induction or deduction. If the formation of science were an exclusively inductive process, the law of the development of the sciences would be one of continuous movement from complexity and particularity to simplicity and generality; if exclusively deductive, the resultant law would be just the opposite, and precisely what Comte supposed it to be, one of uninterrupted advance from the general to the special, from the simple to the complex, from the abstract to the concrete; but the process of scientific discovery being both inductive and deductive, the order of the evolution of the sciences cannot be entirely or continuously in either of the directions indicated, and cannot be either so absolute in itself or so easily ascertainable by us as Comte would have us to believe. In laying down his law of the filiation of the sciences he overlooked all that is empirical and inductive in the sciences, treated each science as if it had been a single truth, and assumed that the order of the succession of the sciences was determined solely by pure deductive reason. In all this he erred most grievously, and simplified his problem most unduly. If science can be built up only by the combined resources alike of induction and deduction, we may be entitled to say, in a general way, that this science must precede that, but not to say, in an absolute way, that this whole science must precede that whole science.

Are we to conclude, then, that Spencer is wholly right and Comte wholly wrong? That is by no means necessary. The association of induction and deduction, of generalisation and specialisation, of analysis and synthesis, in the growth of science, requires us to believe that the sciences spring up together and influence each other to an extent unrecognised by Comte, but not to disbelieve that some sciences are naturally antecedent to others, or even that the sciences of which the phenomena are most general and simple must be further developed than those conversant with phenomena more special and complex. Biology may not only develop simultaneously with Physics and Chemistry, but even suggest to them problems on the solution of which their progress is greatly dependent, while yet all its doctrines must be superficial unless based on the teachings of a Physics and a Chemistry which have attained a relatively high perfection. Although Comte did not see with sufficient clearness to what extent the sciences develop spontaneously and simultaneously, he was not mistaken in so far as he held that one fundamental science does come before another — on the whole, although not wholly — and that in virtue of the relative simplicity, generality, and independence of the laws which they set forth. We may assign full weight to all that is true in the objections urged by Spencer in his criticism of Comte's scheme of filiation of the sciences, and yet reasonably hold that, in the main, it is Comte who is in the right, and that Spencer's view that there is no true order of filiation of the sciences is an exaggerated inference from his facts, and implies that the progress of knowledge is without method or law.

Let us now confine our attention for a moment to the fundamental physical sciences of Comte—Astronomy, Physics, Chemistry, and Physiology. It is obvious, I think, that the first—Astronomy—is not of the same rank as the others. It is not a science of general properties, but of particular objects, which is what no fundamental science is. The fundamental sciences are not classed according to individual objects. Every object is complex, and can only be fully explained by the concurrent application of various sciences. The stars have a mathematics, physics, and chemistry; a mineralogy also, and perhaps a botany and zoology, and conceivably a psychology and sociology. What Comte means by Astronomy is, of course, only the mathematics and physics of the stars; but why, then, make it co-ordinate with the mathematics and physics which include it, or by their synthesis constitute it? The mathematics and physics of the stars would require to be entirely distinct from the mathematics and physics of the earth—*i.e.*, to be no mathematics and physics at all, but things essentially different, before they can be entitled to the place which Comte assigns them — a place separate from all other mathematics and physics. Even if it were true that Astronomy became positive science long before terrestrial physics, this would prove no more than that it was the simplest and most manageable part of physics; it would in nowise prove that it was no part of physics. But the alleged fact cannot be made out. For just as it was impossible to understand the geometrical relations of the celestial bodies while ignorant of the geometrical relations of terrestrial bodies, so it was impossible to apply physical conceptions and generalisations to the stars without having drawn them from our experience of the earth, or at least without applying them at the same time to the earth. The laws of motion, weight, force, &c., which rule in celestial, rule also in terrestrial physics. The great law of gravitation, which regulates the motion of the stars, was, according to the well-known story, suggested to Newton by the fall of an apple, and could certainly not have been ascertained and verified by him if he had been ignorant of the laws of falling bodies, the law of the composition of forces, and the law of centrifugal force, which Galileo and Huygens had previously discovered to rule terrestrial phenomena. We must, therefore, strike out Astronomy from the list of fundamental physical sciences. There then remain only Physics, Chemistry, and Biology. Are these fundamental physical sciences? Are they the only fundamental physical sciences? So far as our present knowledge goes, we must, I believe, answer both questions in the affirmative. These sciences are fundamental, not being able to be resolved into any other sciences or into one another. They are the only fundamental physical sciences because the only irresolvable attributes of matter are physical forces, chemical affinities, and vital properties. Those who make a longer list overlook a distinction without which the whole subject of the relationship of the sciences must be an inextricable imbroglio—the distinction between fundamental and derivative, primary and secondary, simple and complex sciences.

Another French philosopher, contemporaneously with but quite independently of Comte, strenuously occupied his mind during many years on the classification of the sciences, and published, in 1834, the first part of an *Essai sur la Philosophie des Sciences*, the second part of which, completing the work, only appeared in 1843. This philosopher was the illustrious André-Marie Ampère, a man equally remarkable for the extent and the profundity of his knowledge, keenly interested in all the sciences, a brilliant discoverer in several of them, and in particular, as a thoroughly competent authority, the

Ampère.

late Principal Forbes, of St Andrews, observed, "at least as well entitled as any other philosopher who has yet appeared to be called 'the Newton of Electricity.'"

M. Ampère proposes his classification as founded upon a consideration of the sciences themselves. It is, he conceives, in accordance with the conditions of natural classification as exhibited, for example, in Botany. It aims to bring together analogous sciences, and to group them according to their real affinities. It is certainly remarkable for its regularity and symmetry. It proceeds thus: All science has reference to one of two general objects—the material world and thought. This gives rise to the natural division of the sciences into sciences of matter and of thought, or, as Ampère calls them, cosmological and noölogical sciences. Hence all our knowledge is embraced under one or other of two kingdoms. Each kingdom is in its turn the subject of a twofold division. The cosmological sciences separate into those which have for object the inanimate world, and those which occupy themselves with the world of life and organisation, the first of these classes comprehending the mathematical and the physical sciences, and the second the sciences relative to natural history and the medical sciences. In like manner, the sciences of thought divide into two sub-kingdoms, of which the one includes the noölogical sciences properly so called, and the other the social sciences. From these spring, in consequence of another division, four branches, as in the case of the cosmological sciences.

We need not exhibit farther the general scheme. If we confine our attention to the strictly noölogical branch we shall find that it separates into two—the philosophical and the moral sciences. And if we confine ourselves to the moral sciences, we find these also to be two — Ethics and Thelesiology. Then, Ethics, which embraces all that can be known relative to the characters, manners, and moral conduct of man, divides into two parts—Elementary Ethics, which includes Ethography and Physiognomony—and Ethognosy, which comprises Practical Morality and Ethogeny. Thus in Ethics, a science of the first order, there are, according to Ampère, two sciences of the second order and four sciences of the third order.

In the same way Thelesiology, which is conversant with the will, with duty, and the end of man, embraces two sciences of the second order — Elementary Thelesiology and Thelesiognosy — and four of the third order — Thelesiography, Diceology, Apodictic Morality, and Anthropotelic. Thus Moral Science comprehends two sciences of the first order, four of the second order, and eight of the third order. There can be no doubt that this is ingenious, and it is but a very small specimen of the ingenuity of the scheme as a whole. Every science of the first class includes, according to Ampère, four sciences of the third order, and this alleged fact he explains by an alleged law of scientific thought. Intelligence in examining any subject whatever must, he holds, follow a process of four stages. In the first stage, called *autoptic*, it is limited to the simple inspection of its objects; in the second, the *cryptoristic stage*, it investigates their inner and hidden natures; in the third, or *troponomic stage*, it traces the changes which they undergo in time and place, and seeks, from the experience of these changes, to ascertain their laws of change; and in the fourth, or *cryptologic stage*, it occupies itself with what is most uncertain, abstruse, and difficult to discover in their causality and destination. These stages consequently correspond to four epochs of intellectual growth in individual and social development.

The very regularity of the foregoing scheme is an objection to it. Nature is less symmetrical than it represents her to be. She observes order, indeed, and obeys mathematical laws; but she does not incessantly go on dividing by two. She is free and varied in her operations, and none of her secrets of much value will be discovered by so simple a process as a succession of divisions by two. Further division is, in Ampère's scheme, pushed to an excess which tends to defeat the great end of a classification of the sciences. That end is so to group and co-ordinate the sciences that they may be seen together as harmonious parts of a great whole in which the universe is truthfully mirrored. But if we proceed to divide and divide, unsatisfied, as Wordsworth says, "while littleness may yet become more little," we break down all grandeur, destroy all life, and amid the multiplicity of details lose sight of those fundamental laws and relations which are most worthy of our study. If Ampère had divided less he would certainly have succeeded much better in his attempt to form a philosophy of the sciences. The elaborateness of his scheme weighed him down and prevented his rising to a general doctrine exhibiting the unity of science and reflecting the unity of the universe. He found that even in two volumes he could do no more than give a general idea of each of the multitude of sciences to which he assigned a place, although aware that an exposition of the fundamental truths and general methods of science is essential in a philosophy of the sciences. If the trees did not hide from himself the forest, they certainly prevented him from describing it to others.

The scheme under consideration has, however, even greater defects than those just indicated. One is that it makes no distinction between arts and sciences, but treats the former as of the same nature and as to be placed on the same level and ranked in the same line with the latter. It represents, for example, Technology as following Chemistry and preceding Natural History, and often thus puts arts and sciences side by side. This ignoring of the true relationship of science and art—this confounding of knowledge and its application, of the *quid* and *quid lucri*—renders a true classification of the sciences absolutely impossible. And it involves another error as great as itself—the ignoring of the distinction between fundamental and derivative sciences. Unless the arts are separated from the sciences the sciences themselves cannot be distributed into fundamental and derivative. To set aside the distinction between dependent and independent at one point of the scheme is to necessitate its being set aside throughout.

It would not be difficult to show that Ampère's sciences of the third order are seldom natural divisions of his sciences of the first order. In fact, the very conception of there being in each science of the first order four sciences of the third order corresponding to four distinct points of view from which their common subject may be studied is illusive. Even conceding the four points of view, it cannot be reasonably held that there are separate sciences to correspond to them. The points of view represent only stages of the scientific process; they are only the series of steps by which science is attained. Science corresponds to the process as a whole, not to any particular point or stage of it. Science may well begin with the simple inspection of objects, and must, of course, end with their full comprehension; but this is not the slightest reason for supposing, as Ampère does, that there are sciences of simple inspection and sciences of full comprehension—autoptic sciences and cryptologic sciences. With all his knowledge and ingenuity Ampère failed to classify the sciences aright, and still more to found a philosophy of the sciences. His work, however, is most instructive, and not unworthy even of his great reputation.

The celebrated socialist, P. J. Proudhon, published in 1843 a work entitled *De la Création de l'Ordre dans l'Humanité*, in which traces of the influence both of Comte and Ampère are deeply marked. Comte's law of three states is unqualifiedly adopted in substance, although the terms in which it is expressed are changed, metaphysic being called philosophy or sophistic, and the doctrine of the sciences or positive philosophy being designated metaphysic, so that in Proudhon's phraseology the Comtist law runs thus: "Religion, philosophy, science; faith, sophistic, and method (metaphysic) Proudhon.

—such are the three moments of knowledge, the three epochs of the education of the human race." Ampère's classification of the sciences is expounded and highly commended. At the same time, it is held not to be the absolute or only true classification of them. "The mind," says Proudhon, "may find in Nature a multitude of systems, according to the point of view which it takes up, although Nature herself follows none of them exclusively." He undertakes, in particular, to show that for the *quaternary* distribution of Ampère a *ternary* may be substituted, not less natural, regular, and precise. Of this ternary classification he would make the ordinary distribution into kingdoms in Natural History — *mineral, vegetable,* and *animal*—the basis, and then would divide the sciences, according as they are descriptive or declarative of phenomena, or as they study forces, motions, progress, changes, or as they formulate laws and determine relations. In other words, he would reduce Ampère's four points or stages of the scientific process to three, but retain his vicious principle of regarding such mere points or stages as the roots of distinct sciences. He has not exhibited his ternary classification in detail, but he professes to have worked through the whole scheme of Ampère, changing it everywhere from quaternary to ternary —"absolutely as if I had transcribed our decimal arithmetic into a duodecimal arithmetic." This I can readily believe, although I would infer from it not, as Proudhon does, that both systems are alike natural, but that both are alike arbitrary.

Duval-Jouve.

J. Duval-Jouve in his *Traité de Logique, ou Essai sur la Théorie de la Science*, 1844, has dealt with the classification of the sciences in pp. 374-393. He distributes them into cosmological and noölogical sciences, and subdivides the former into mathematical and physical classes. That is manifestly insufficient. The work, however, can be safely commended for its judicious counsels as to the study of the sciences of reasoning and of physical and psychological observation.

Rosmini.

Two Italian philosophers of rare genius, and whose influence on the thought and life of their nation was great and salutary,—Antonio Rosmini and Vincenzo Gioberti,—now claim our attention; but, of course, only in so far as they have dealt with the special problem which at present concerns us. Neither dealt with it as an independent problem, only to be solved by a comparative study of the sciences themselves; on the contrary, both professedly evolved their classification of the sciences from the fundamental principle of their philosophies. That seems to me an altogether illegitimate procedure, resting on an assumption as to the relation of philosophy to the sciences just the reverse of the truth.

Rosmini's views on the classification of the sciences are to be found in his *New Essay on the Origin of Ideas* (first edition, 1830), *Philosophical System* (first edition, 1845), and *Logic* (first edition, 1854), the first two of which have appeared in English. On the ground that every cognition must have matter and form, he represented the sciences as primarily divisible into *material* and *formal;* and on the ground that the form of cognition is at once the source of all intelligence and alone knowable *per se,* he held that the science of the form must precede all other sciences and supply the principle of their encyclopædic arrangement. This first science, which he called Ideology, he regarded as the only *pure science,* all other sciences being in relation to it only *applied sciences.* But he was not content merely with this division, and so proposed another corresponding to the aspects of Being, that one necessary and objective form of intelligence to which he believed all the other forms of cognition could be reduced, and also to the modes of mental activity by which these aspects of Being are apprehended. Thus, viewing Being as *ideal, real,* and *moral,* and intelligence as possessed of *intuition, perception,* and *reason,* he classified the sciences in the following threefold manner:

1. Sciences of intuition, which treat of the ideal and include Ideology and Logic; 2. Sciences of perception, which treat of the real and comprehend Psychology and Cosmology; and, 3. Sciences of reasoning, which treat of what is only discoverable through inference and may be subdivided into Ontological and Deontological Sciences. The Mathematical Sciences have no place in the scheme, nor even the Physical Sciences, the Rosminian Cosmology being only a department of Metaphysics. The Ontological Sciences are said to be Ontology, properly so called, and Natural Theology. The Deontological Sciences are those which treat of the perfection of being, and of the way in which this perfection may be acquired and lost; and as they are distributed in a somewhat minute and decidedly artificial manner, it may suffice to say that they comprehend not only Moral sciences usually so called, but Æsthetic sciences, Political sciences, Pædagogics, and Economy. Language and history are not represented as the special objects of distinct sciences, but a scientific study of history is recognised to be an important means of advancing the Philosophy of Politics.

To Gioberti the first principle of Rosmini seemed a vain abstraction and his method essentially false; and he resolved for his own part to start not with **Gioberti.**

indeterminate ideal being, but with an object at once ideal and real, and to evolve what it implied, not by a psychological method, which can only reflect the mind's attention to itself, but by the only true philosophical method, the ontological, which reflects the nature and manifestation of the object. He deemed that he found at once a point of departure and a law of procedure satisfying the requirements of the case in a synthetic judgment, comprehensive of all being and knowledge, for which the appropriate expression is the ideal formula *Ens creat existentias — Being creates existences.* In his *Introduzione alle studio della Filosofia* (vol. iii. cap. v.), published in 1840, he has explained how the sciences may be arranged in accordance with his formula. It is by a method which, if not eminently satisfactory, is at least eminently easy. The ideal formula is itself the "*suprema formula enciclopedica,*" and all sciences, it is held, may be directly referred to one or other of its terms. The subject (Being) is the theme of Philosophy Proper, which includes the sciences of Ontology and Theology. The copula (Creates) yields the sciences which are concerned with the relationship of Being to Existences and of Existences to Being, the relationship of Being to Existences being treated of by the Science of time and space (Mathematics), and the relationship of Existences to Being by the Sciences of the true, the good, and the beautiful (Logic, Ethic, and Æsthetics). The predicates (Existences) originate the sciences which are conversant with the effects or results of the creative act—namely, Psychology, Cosmology, and the various special Physical Sciences. Besides these Rational Sciences there are Super-Rational Sciences based on revelation; they are, however, to be classified in the same manner as the Rational Sciences. Such is the scheme of classification proposed by Gioberti. It has various obvious faults, but these it seems unnecessary to specify, seeing that the foundation of the whole scheme is utterly untrustworthy. The "ideal formula," on which everything is made to depend, is admittedly the expression of an act of mystic intuition, and really an arbitrary affirmation.

V. FROM WHEWELL TO ZELLER.

Dr William Whewell, a man of extraordinary versatility, industry, and knowledge, published in 1837 a *History of the Inductive Sciences,* and in 1840 a *Philosophy of the Inductive Sciences.* In the latter he treated of the classification of the sciences. The work was greatly altered, even in the arrangement of its parts, in the third edition, where the discussion of the problem and the classi- **Whewell.**

fication proposed will be found in the volume entitled *Novum Organon Renovatum*, B. II. ch. ix. "The classification depends neither upon the faculties of the mind to which the separate parts of our knowledge owe their origin, nor upon the objects which each science contemplates, but upon a more natural and fundamental element—namely, the *Ideas* which each science involves. The Ideas regulate and connect the facts, and are the foundations of the reasoning, in each science." It is not necessary, Dr Whewell observes, that the Idea on which a science is founded should be an absolutely ultimate principle of thought, or that it should be the only Idea involved in the science. "Each science may involve, not only the Ideas or Conceptions which are placed opposite to it in the list, but also all which precede it." Whewell's groups of sciences are as follows: 1. Pure Mathematical Sciences, including Geometry, Arithmetic, Algebra, and Differentials, and based on the ideas of space, time, number, sign, and limit. 2. Pure Motional Sciences, including Pure Mechanism and Formal Astronomy, and based on the idea of motion. 3. Mechanical Sciences, including Statics, Dynamics, Hydrostatics, Hydrodynamics, and Physical Astronomy, based on the ideas of force, matter, inertia, and fluid pressure, which are modifications of the idea of cause. 4. Secondary Mechanical

Sciences, including Acoustics, Optics, Thermotics, and Atmology, and based on the ideas of outness, medium of sensation, intensity of qualities, and scales of qualities. 5. Analytico-Mechanical Sciences, including Electricity, Magnetism, and Galvanism, and based on the idea of polarity. This group and the immediately preceding one may, it is indicated, be brought into connection as constituting the two branches of Physics. 6. Analytical Science, identified with Chemistry, and held to correspond with the ideas of element, chemical affinity, and substance or atoms. 7. The Analytico-Classificatory Sciences — namely, Crystallography and Systematic Mineralogy, which have symmetry and likeness for ideas. 8. The Classificatory Sciences—namely, Systematic Botany, Systematic Zoology, and Comparative Anatomy, which have as their ideas degrees of likeness and natural affinity. 9. The Organical Sciences, or Biology, founded on the ideas of vital power, assimilation, irritability, organisation, and final cause. 10. Metaphysics, coincident with Psychology, and corresponding to the ideas emotion and thought. 11. The Palætiological Sciences, comprehending Geology, Distribution of plants and animals, Glossology, and Ethnography, and springing from the idea of historical causation. And, 12. Natural Theology, which rests on the idea of a first cause.

In examining this scheme its fundamental assumption, that each science presupposes a special *a priori* idea, is by no means found to be borne out. Reasons to doubt its truth soon present themselves. Suspicion thereof is forced on us by Dr Whewell himself, even in regard to the mathematical sciences. Algebra, for example, rests, he tells us, on the *a priori* idea of *sign.* But is *Sign* an *a priori* idea? And if so, will it not be difficult to discover any *a posteriori* idea? Naturally, however, as soon as Dr Whewell passed beyond the province of mathematics his difficulties greatly increased; and, in fact, with every forward step he took the ineptness and inapplicability of the principle he had assumed were made more manifest. He soon reached sciences which he had to refer to things never heard of before as *a priori* or fundamental *ideas—e.g.*, fluid pressure, medium of sensation, intensity of qualities, polarity, atoms, &c. The mental sciences he wisely refrained from attempting to subdivide or trace to root ideas. There are other serious defects in Whewell's scheme. Thus, Mechanical Science and Analytical or Chemical Science have no higher rank assigned them than Secondary Mechanical Science and Analytico-Mechanical Science; that is to say, they are put on a level with sciences which are only branches or applications or combinations of themselves.

Then there follow as Analytico-Classificatory and Classificatory Sciences what are simply the unscientific parts of Mineralogy, Botany, and Physiology. Observation, classification, and description of phenomena are not science, although they necessarily precede it. Anatomy, for instance, regarded merely as descriptive, is a subordinate science; it is the series of observations and classifications preparatory to the science of Physiology; it is no more a complete science than would be a description of the lines and figures employed in Geometry. Then, tracing the scheme a little farther, we find Metaphysics identified with Psychology, which in reality amounts to the entire elimination of Metaphysics; and Geology and the Science of the distribution of plants and animals appearing, as Palætiological Sciences, *after* Metaphysics or Psychology, quite separated from Mineralogy, Botany, and Zoology, with which one would naturally have expected them conjoined, and with which they are certainly in much closer connection than with Metaphysics or Psychology. It savours of the ludicrous to represent Natural Theology as in closer contact with the Palæontological Sciences than with any others, on the ground that they are conversant with historical causes and it with the first cause. There is, finally, an objection of wider sweep which I have not time to work out. Whewell fixes the order of the sciences by referring them to what he deems their corresponding ideas. But how has he determined the order of the ideas? And has he determined it aright? It would be easy to show that he arranged them in a haphazard way, with extremely little regard to their rational connections.

Lubbock. The *Remarks on the Classification of the Different Branches of Human Knowledge*, published in 1838 by J. W. Lubbock, possess hardly any value. The general division of the classification recommended is into History, Philosophy, and Fine Arts; and Philosophy is subdivided into Religion, Jurisprudence, Intellectual, Moral, and Political Philosophy, Logic, Mathematics, Natural Philosophy, Natural History, Medicine, and Arts, Trade, and Manufactures.

Lindsay. In *Progression by Antagonism*, published by the late Earl of Crawford (when Lord Lindsay) in 1846, a "classification of human thought" is put forth based on the general theory of development expounded in that exceedingly interesting book. While the admission is made that no art or science springs from imagination alone or reason alone, it is also held that each art or science must be distinguished by and classed under the predominant faculty which originates it. Spirit ruling sense predominantly by imagination gives rise to Symbolism, Fine Arts, Rhetoric, Poetry, and History; ruling predominantly by reason to Science, which is Speculative or Pure and Practical or Applied, both subdivisible into Physical and Metaphysical, each of which contains many separate sciences; and ruling by reason and imagination in harmonious co-operation to Philosophy, also to be distributed into Speculative and Practical. The order of classification is said to be "determinable by that in which the individual, national, and universal mind applies itself to the respective arts and sciences."

Ramsay.

In 1847 the late Sir George Ramsay published *A Classification of the Sciences, in Six Tables.* The primary division is into: 1. Mental Sciences; 2. Physical Sciences; and 3. Mathematics. It does not seem to have occurred to the author that, even if these were the chief classes of the sciences, the order in which they are arranged is the reverse of natural. Theology finds a place only under Moral Philosophy, one of the mental sciences. The grouping is altogether of an external and unphilosophical kind.

Schopenhauer.

In 1844 Schopenhauer, in the second edition of his chief work, *Die Welt als Wille und Vorstellung,* proposed a scheme of distribution of the sciences which, perhaps, deserves to be noted only as an ingenious curiosity. Schopenhauer, it has been said, accepted one of Kant's categories, and threw the other eleven out of the window. It is on his one working principle of the sufficient reason that he hangs his scheme of classification. Every science is regarded as exemplifying predominantly one of the forms of that principle. The main division is into I. Pure *a priori* sciences, and II. Empirical or *a posteriori* sciences. The former are subdivided into (1) the doctrine of the principle of Being in (*a*) Space—Geometry, and in (*b*) Time—Arithmetic and Algebra; and (2) the doctrine of the principle of knowledge—Logic. The latter are concerned with the principle of becoming, or law of causality, and in its three forms of *cause, stimulus, and motive.* Hence they are grouped as follows: (1) The doctrine of causes. (*a*) General: Mechanics, Hydrodynamics, Physics, Chemistry. (*b*) Special: Astronomy, Mineralogy, Geology, Technology, Pharmacy. (2) The doctrine of stimuli. (*a*) General: Vegetable and Animal Physiology, with Anatomy as auxiliary science. (*b*) Special: Botany, Zoology, Comparative Physiology, Pathology, Therapeutics. (3) The doctrine of motives. (*a*) General: Ethics, Psychology. (*b*) Special: Jurisprudence, History.

P. E. Dove. Patrick Edward Dove, in his *Theory of Human Progression* (1850),—published at first anonymously, but afterward acknowledged,—treated the problem under consideration with great clearness and vigour.

The general aim of his work was to show the natural probability of the coming of a reign of justice,—the advent of a moral millennium,—and, as essential to this, to prove that there is a natural progression of the mind in the extension of its knowledge and the improvement of its practice. Such a progression implied, according to Mr Dove, the consecutive evolution of the sciences and their logical dependence on each other. The classification which he proposed rested on the principle that every science must have a distinctive object-noun, the place of which among the categories of the mind determines the place of the science among the series of the sciences. The object-noun of a science is the primary condition of its existence, and of the forms of that noun the science exclusively treats. The connection of object-nouns is such that the sciences follow in a determinate order, the one in which they must necessarily be studied and also that in which they must necessarily be discovered. It is an order of ever-increasing complexity, each later science including not only its own distinctive concept, but those of all the sciences which precede it. Thus, Logic is the first and simplest science. Arithmetic is nothing more than Logic applied to number. Algebra is Logic and Arithmetic applied to quantity. Geometry (in its larger sense) is Logic, Arithmetic, and Algebra applied to space. Statics is Logic, Arithmetic, Algebra, and Geometry applied to force. And if we look at the object-nouns or substantive concepts we shall find, according to Mr Dove, that Logic has two branches, the one treating of identity and the other of equality; that Arithmetic adds to identity and equality number; Algebra to identity, equality, and number quantity; Geometry to identity, equality, number, and quantity space; and Statics to all these forces. "In this order," we are told, "the mathematical sciences must necessarily be classed, and in this order the mathematical sciences must necessarily be discovered. Ten thousand men originating the mathematical sciences by a process of independent investigation would necessarily discover them in this order; and were ten thousand worlds peopled with human beings to go through the process of making anew the mathematical sciences, every one of those human races would pass through the same intellectual course, and evolve the abstract sciences exactly in the same necessary order. The constitution of human reason forbids that it should be otherwise, one science being impossible until its antecedent is so well known as to be capable of subjective operation. Thus, unless the laws of identity are known, there can be no investigation of the laws of equality; and until the laws of equality are known, there can be no investigation of the laws of number; and until Arithmetic is known, there can be no investigation of the laws of quantity; and until the laws of quantity are known, there can be no investigation into the relations of space; and until Geometry is known, there can be no Statics."

The sciences which have just been mentioned — the mathematical sciences — are all devoid of any idea derived from sense. When, however, they are applied to the substantives and operations of real life, they originate another order of sciences — the physical sciences — which arise one after another in similar order of complexity. The first and simplest of these sciences is Dynamics, which is closely connected with the last of the mathematical sciences — Statics — Statics dealing with forces which neutralise each other, and Dynamics with forces which produce motion, the simplest and most universal function of matter. It is by adding to motion one physical characteristic after another that the physical sciences are consecutively evolved. Thus, add to it weight or resistance, the next most general property, and you have Mechanics; add still further sound, light, and heat, and you have as corresponding sciences Acoustics, Optics, and Thermology; add again magnetic force, electric force, and affinity, and you have the sciences of Magnetism, Electricity, and Chemistry; and these three sciences are, in their turn, the necessary preparations for a new, a third order of sciences—the sciences of organisation—comprehending Vegetable and Animal Physiology. These again lead to another, a fourth order of sciences, the man-sciences, or sciences of human action, which are a sensational and inductive science called Political Economy and conversant with utility, and a moral and deductive science called Politics and conversant with equity. The last science is Theology. It closes and completes the book of science properly so called. But beyond science lies Critical Philosophy. Science is direct and spontaneous, and seeks only to determine what is true in that which it makes its object, whether mind or matter; whereas Philosophy is subjective and reflective, and inquires not into the truth of thought, but into its form and mechanism, endeavouring with the whole mass received from the whole circle of the sciences to read aright the phenomenon of knowledge.

The scheme of Dove, it will have been remarked, has an obvious resemblance to that of Whewell. It proceeds throughout on the same assumption, although that assumption is applied with much greater tact and plausibility by Dove than by Whewell. These two objections may be urged against the scheme as a whole: 1. The conception of object-nouns on which it rests is erroneous. An object-noun is implied to be something very different from what is ordinarily meant by the object of a science—namely, that about which the science is conversant; it is supposed to be some single idea the application of which to appropriate objects constitutes science. But it is only of the purely abstract sciences that this can be with any appearance even of truth maintained. Inductive science at least originates in no such way; it needs only an object in the sense of a certain kind of material subject to laws discoverable by the inductive process. 2. More even than the scheme of Comte that of Dove is vitiated by the hypothesis that the order of the formation of the sciences is absolutely fixed and necessary, proceeding on one straight line, and incapable of being other than it is. Comte only makes his scheme exclusively rational and deductive in the working of it out and by taking no account of induction as counteractive of deduction, whereas Dove lays down *a priori* principles and a deductive procedure as the very groundwork of his whole system. In so doing he builds upon the sand. Reason shows that the order of the formation of the sciences must be different from what he affirms it to be. Facts prove that it is different. Spencer, in his masterly criticism of the Comtist classification, has decisively established the truth of both of these affirmations.

Let us now look a little at the details of Mr Dove's scheme. Logic is placed at the head of the sciences; it is said to be the first, because the simplest, of the sciences. But simplest in this case means most abstract, and the most abstract, instead of being always first, is generally last. That Logic is more abstract than Arithmetic, Algebra, and Geometry, instead of being a conclusive reason for supposing it to be in the order of study and discovery before them, is a reason for suspecting it to be behind them. And, in fact, both Arithmetic and Geometry preceded it. If it be said there can be no reasoning in number or space which does not presuppose identity and equality, the answer is twofold, for, first, in a more relevant sense identity and equality presuppose number, space, and other concepts regarded as later, since one thing is not identical with or equal to another unless identical or equal in number, space, &c.; and, secondly, it needs no science to give us the notions of identity or equality before we can proceed to study any other science, as these notions are firmly and operatively in our minds before all science.

Then, further, why confine Logic to reasoning in identity and equality? Why not extend it to all reasoning? It will be said, because Arithmetic is conversant with reasoning in numbers, Algebra with reasoning in quantity, Geometry with reasoning in space, &c. But no; these sciences are conversant with number, quantity, space, and their laws, while reasoning and its laws in number, quantity, space, or any other concept or matter, are the object of Logic, which is therefore not, strictly speaking, before any science, but pervasive of all science, having to trace the connective tissue of all knowledge, the forms and methods of all sciences. This view of it, however, would have quite deranged Mr Dove's serial arrangement. It leaves, likewise, no place for his philosophy; for, according to him, it is Philosophy which has to do with the form and method of thought. If, therefore, he had taken a sufficiently comprehensive view of Logic he would have seen that it included and fulfilled all the functions which he assigned to Philosophy.

As to the sciences grouped as Mathematical —Arithmetic, Algebra, Geometry, and Statics—it is obvious to remark that important mathematical sciences are entirely omitted; and that to describe Algebra as Logic and Arithmetic applied to quantity, or Geometry as Logic, Arithmetic, and Algebra applied to space, conveys no meaning, and cannot be asserted to be erroneous only because unintelligible. Dove represents Arithmetic as the first of the mathematical sciences, whereas Whewell, it will be remembered, assigns that honour to Geometry. Both are, perhaps, right and both wrong. Either science may have originated before the other, or they may have been of simultaneous origin and growth. Statics is not a mathematical science at all. Instead of force being, as Dove supposes, a mathematical conception and motion, the first and simplest of physical conceptions, it is motion which is the mathematical and force which is the physical conception. There is a science of pure motion, the science now generally called Kinematics; and it is a mathematical science, not only because it treats of motion, displacement, and deformature, tortuosity, and curvature, altogether independently of force, mass, elasticity, temperature, magnetism, electricity, which are all physical attributes, the first not less than the last. The arrangement of the Physical Sciences is also defective. In particular, secondary sciences are put on a level or equality of rank with those of which they are branches, or at least from which they are derived. Passing from the Physical Sciences, Psychology is found to have been omitted altogether, although it must be regarded as the very foundation of the so-called Man-Sciences. There can be no science of human actions if there be none of human nature. Yet Psychology is not merely a Man-Science. There is a Comparative Psychology as certainly as there is a Comparative Physiology.

The sphere of Psychology includes every fact of sensation, every form of consciousness, animal as well as human; it has to do with the psychical life of all sentient creatures from the animalcule to the man. This of itself shows that it must always be arbitrary to make an exclusive instead of an inclusive group of Man-Sciences. In his *Theory of Human Progression*, Dove ranked Politics or Ethics as one of the Man-Sciences, erroneously identifying Politics and Ethics. It is curious to observe how, in his *Elements of Political Science*, published four years later, he gave Ethics or Politics a quite different position. He placed it, in this latter treatise, as an abstract science immediately after Statics; in other words, he ranked it as a mathematical science, and held that, owing to the *ideal* character of its truths, it stands on a higher level than the mental or other inductive sciences.

M. Cournot, a man of remarkable capacity both for philosophical speculation and scientific research, treated of the co-ordination of the departments of human knowledge in his *Essai sur les fondements de nos connaissances* (tom. ii. ch. xx.-xxii.), published in 1851. He followed to some extent Bacon, and to a much larger extent Ampère, although he also criticised both with characteristic acuteness and independence. He may be said to have adopted, **Cournot.**

in the main, Ampère's classification, but with numerous and important modifications which are mostly decided improvements. He rejected "the artifice of bifurcation." Instead of commingling and confounding, as Ampère did, arts and sciences, he entirely separated them. He attempted to distinguish carefully between science strictly so called and history, and founded on the distinction a division of the sciences into two great series—namely, (*a*) a cosmological and historical series and (*b*) a theoretical series. There are thus three parallel series of the kinds or divisions of knowledge—a technical series, a cosmological and historical series, and a theoretical series. Our author did not apply the distinction between science and philosophy, like that between science and history, as a principle of classification. For that his reason was that philosophy cannot be sharply separated from science, while history can. Philosophy, he held, has no special object of its own; is not a science or group or series of sciences; but is an indispensable element of all sciences; lies at their root, pervades their ramifications, and reaches to their summits. The series of theoretical sciences he divided into five groups—the mathematical, physical, biological, noölogical, and political sciences. Psychology he placed among the biological, not the noölogical sciences; on the other hand, he regarded Natural Theology as a noölogical science. Few of his groups seem to comprehend just the sciences which they ought to contain; but the distribution as a whole has very great merits.

An American author, Prof. W. D. Wilson, published in New York, in 1856, an *Elementary Treatise of Logic*, which contains, in its last chapter, a classification of both Sciences and Arts. They are divided into three classes—namely, Theoretical Sciences, Practical Sciences, and Productive Arts; so that the scheme is essentially a modernised reproduction of the Aristotelian distribution of philosophy. Each of the three classes, we are told, "naturally divides itself into two departments, differing in the *first* class, both in the starting-point and in the method; in the *second* class they differ in the starting-point only; and in the *third* class the two departments differ chiefly in the object in view—the one producing objects of beauty, and the other objects of utility." The departments of the Theoretical Sciences are: 1. Exact Sciences, and, 2. Pure Sciences. The former includes Meteorology, Ouranography, Geology, Geography, Chemistry, Mineralogy, Anatomy, Physiology, Botany, Zoology, Ethnology, Psychology, and History; the latter Arithmetic, Geometry, Algebra, Calculus, Trigonometry, Analytic Geometry, Analytics, Method, and Ontology. The departments of the Practical Sciences

Wilson.

are: 1. Mixed Sciences, and, 2. Ethical Sciences. The former comprehends Mechanics, Astronomy, Hydrostatics, Hydraulics, Pneumatics, Acoustics, and Optics; the latter Ethics, Polity, Natural Religion, Jurisprudence, Church Polity, and Revealed Religion. The departments of the Productive Arts are: 1. Fine Arts, and 2, Useful Arts. The former contains Gardening, Architecture, Sculpture, Painting, Music, and Poetry; and the latter Agriculture, Metallurgy, Technology, Typography, Engraving, Commerce, Medicine, Rhetoric, Political Economy, and War. This scheme is much inferior to that of Cournot. It is impossible to regard the order in which the sciences are arranged in it as the order in which they have been discovered, or that in which they should be studied, or as a natural order of any kind. A number of the so-called Exact Sciences are obviously and necessarily less exact than the so-called Pure Sciences and Mixed Sciences. The designation Exact Sciences is an infelicitous one, as all science is only science on condition of being exact.

There is nothing on our subject worth mentioning in the hazy and confused *Organismus der Wissenschaft* which Adolf Helfferich published in 1856. Science he defines as "the rational or ideal reproduction of the real human personality," and, therefore, holds that "the organism (*Gliedbau*) of science must correspond to the organisation (*Gliederung*) of the human being."

In Sir Wm. Hamilton's *Lectures on Metaphysics*, **Hamilton.** published in 1859, but delivered from 1836-37, there is a classification (see Lect. VII.) which, although comprehending only the mental sciences, may be noticed here, because if good for the mental sciences it should be equally good for the physical sciences. On the other hand, if no physical philosopher would think of arranging the sciences with which he is conversant as referring to the facts, the laws, and the results of the material world, or, in other words, as phænomenological, nomological, and ontological; if, on the contrary, he must recognise that such an arrangement would contravene every true notion of what science is, it may be inferred that such an arrangement of the mental sciences cannot be more tenable, less unscientific, less destructive of every true notion of the nature of science. Let us consider, however, Hamilton's classification in itself.

He starts from the common but erroneons notion that philosophy is equivalent to mental science. Then he proceeds to divide and distribute philosophy thus understood on the supposition that mind or consciousness yields us facts, laws, and results. If we deal merely with the facts or phenomena of mind, we have a mental science or department of mental science which may be called the Phænomenology of Mind, but is generally known as Psychology. Its divisions correspond to the classes of mental phenomena—cognitions, feelings, conative powers. If we deal with the laws of mind we have a Nomology of Mind, Nomological Psychology, which comprises within itself three different Nomologies—one of cognition, Logic; one of feeling, Æsthetics; and one of conation, Practical Philosophy, or Ethics and Politics. If we deal with the results or inferences which the facts of mind or consciousness warrant, we have Ontology, Metaphysics Proper, Inferential Psychology.

Such is the classification of Sir Wm. Hamilton. None of its divisions, major or minor, seem to me correctly drawn.

Begin with the first, the Phænomenology of Mind, erroneously identified with Psychology. What sort of science can that be which deals only with facts or phenomena, which deals with them to the exclusion of laws? There can be no science where there are no laws. Science consists in the knowledge of laws. A mere phænomenology, either of matter or mind, however extensive, however exhaustive, can have no title to be deemed science. Psychology is not such a phænomenology of mind, just because it labours to discover the laws of mind, yea, the most hidden, the essential, and ultimate laws of mind. The separation of facts and laws in science—the assigning of facts to one science and of laws to another—involves not only a false division of the sciences, but the mutilation and destruction of the very idea and life of science, since science is essentially the union of facts and laws, the explanation of facts by laws.

As to the particular divisions of the Nomology, not one of them seems accurately drawn. How can Logic, for example, be called a Nomology of the cognitive powers? On no reasonable view of it, and not even on Sir Wm. Hamilton's own view of it. Logic he held to be the science of the formal laws of thought, and by thought he meant only what is strictly termed discursive thought. In other words, he regarded and treated it as *the science of some of the laws of one of the processes of one of the cognitive faculties*, yet in his scheme of classification represented it to be *the science of all the laws of all the processes of all the cognitive faculties.*

Æsthetics and Ethics are both only in part psychological. The distinctive objects and principles of both can no more be evolved out of any psychological process than out of any physiological or other physical process. And, on the other hand, the properly psychological province of Æsthetics is not inclusive of all the laws of feeling, and yet comprehensive of laws of perception, imagination, and reason; and the psychological provinces of Ethics and Politics are neither limited to nor everywhere as extended as the reign of the laws of action.

Then the Inferential Psychology of Sir Wm. Hamilton does not seem to answer to Metaphysics Proper. Metaphysics is not usually conceived of as a science of results, but as a science of principles. It is almost universally supposed to be occupied with the conditions of all science, which is a very different thing from consisting of the inferences from a particular science. There is a science which deals with the results of all other sciences—a science to which the ultimate conclusions of every science are data from which it draws its own inferences. That science is Natural Theology. When the scientific specialist has reached his highest generalisations, the theologian receives them from him, and, by showing that they are to be regarded as expressions of the manifestation of Godhead, surrounds them with a halo of Divine glory. Metaphysics is quite a different science, being conversant not with what thus overlies, but with what underlies our knowledge of contingent things. Hence Sir Wm. Hamilton's description of Metaphysics answers not at all to Metaphysics, but slightly to Natural Theology. And it will be observed that both the instances which he gives as specimens of the inferences with which Metaphysics is concerned are not metaphysical, but theological truths—the existence of God and the immortality of the soul. But while Hamilton's Metaphysics answers slightly to Natural Theology, it is only slightly and badly, seeing that the truths of Natural Theology ought to be drawn from the results not of psychological science alone, but of all science. All things tell us of God. The mind, indeed, always draws the inference which relates to Him, but it does not always draw it from itself. Further, Sir Wm. Hamilton's Inferential Psychology, as described by himself, is not a psychological science, is not a division of Psychology. Its inferences relate to realities beyond the mind, while explanatory of mind; its truths are reached *through* truths of Psychology, but are not truths *of* Psychology. Sir Wm. Hamilton's classification, in fact, is erroneous from beginning to end—erroneous in its root and in all its ramifications.

The late M. Charles Renouvier, a vigorous and acute thinker who developed and applied the doctrine of phenomenalism with a comprehensiveness and consistency probably unequalled, dealt with the subject of the rational classification of the sciences in the second of his *Essais de Critique Générale*, **Renouvier.**

the *Traité de Psychologie Rationnelle* (ch. xviii.), first published in 1859.[1] He entirely rejected Comte's hypothesis of a hierarchy of the sciences, and regarded their classification as purely a question of logical arrangement. He describes what he calls General Criticism (*La Critique Générale*) as "the common trunk of all the sciences." It has to analyse the universal conditions of knowledge; to study the general nature and laws of experience; and to treat specially of the categories of relation, personality, causality, and finality. From this trunk spring two great branches of sciences—the logical and the physical—which differ not only in their objects, but also in their methods, the logical sciences following the method of ratiocination, and the physical sciences the method of observation and experimentation. The logical sciences comprehend (*a*) logical sciences in the narrower sense of the word, those occupied with the relations of quality—namely, Logic and General Grammar; and (*b*) mathematical sciences, those occupied with the categories of number, position, succession, and change—namely, Arithmetic, Algebra, Mathematical Analysis, Geometry, Rational Mechanics, and Applied Mathematics. The physical sciences include a group of Natural History Sciences (Cosmology, Geology, Mineralogy, Botany, Zoology,

[1] I have seen only the second edition, which is of 1875.

and their subsidiary disciplines), Physics (with Astronomy appended), Chemistry, and Biology. The main line of demarcation among the physical sciences is that which has strictly physical science and chemical science on the one side, and biological science on the other, just as the great division of their objects is into inorganic and organic. There are, however, according to Renouvier, a number of other studies which are not yet definitively separated from philosophical speculation and constituted distinct sciences. These, therefore, he would not class as sciences, but regard as belonging to General Criticism. They include History, Morals, Politics, and Political Economy, and were they sufficiently advanced to be accounted sciences might be classed as Moral Sciences. The tree of science would then have three, not two, great branches.

The foregoing scheme has, I think, serious defects. One is the non-recognition of theological science. It is due, doubtless, to the thoroughness and consistency of M. Renouvier's phenomenalism; but it also indicates that an exclusive phenomenalism is not the whole truth. Then, what M. Renouvier calls General Criticism seems an incoherent and incongruous combination of philosophy and special science. It is identified both with the knowledge which transcends special science because of its universality, and with that which falls below it because of its lack of certainty. That is not a view to be commended. Philosophy should keep to the universal, and cannot be too critical; and it shows itself forgetful of both requirements when it identifies itself with special studies on the ground that they are somewhat too conjectural and uncritical to be deemed sciences. It may, further, be reasonably objected that the conditions of thought and their relations ought to be regarded as the objects, not of *La Critique Générale*, but of a special science with a perfectly definite sphere —a science closely akin to, if not inclusive of, Logic, which treats of the conditions of a kind of thought, discursive thought; also, that Logic has to do with reasoning in quantity as well as in quality, and, indeed, with reasoning in all categories and under all forms. A glance at the order in which the physical sciences are arranged will suggest that Comte's view of "a hierarchy of the sciences" cannot be so wholly false as M. Renouvier contends. Were it not on the whole a natural and true view he would hardly be found conforming to it so much, even when condemning it. It is, likewise, certainly a serious defect in the scheme that so many sciences are left unclassed and unarranged. Notwithstanding his great ability, therefore, M. Renouvier was not in this instance quite successful.

In a tractate entitled *Nuovo Albero Enciclopedico*, published at Naples in 1863, Melchiore Peccenini, of Ferrara, has classified the sciences on the hypothesis that the three chief endowments of mind are *the intellect, the will, and the æsthetic sentiment*, and that the objects which respectively correspond to them are *truth, goodness, and beauty*. Truth, goodness, and beauty are naturally and closely connected, and equally so are all the sciences and fine arts, seeing that they originate in these innate ideas. Common to all the sciences and arts is *being* (*l'ente*), which in relation to intellect is truth, in relation to will goodness, and in relation to æsthetic sense and imagination beauty. Hence, under the head of "Truth (Intellect)" are placed all the sciences which "regard *being* purely with reference to intelligence." Thus, *abstract being* is said to be the object of Ontology or Protology; *concrete being in God* of Natural and Revealed Theology; *concrete being in the soul* of Psychology, Ideology, Logic, Grammar, and Somatics; and *concrete being in matter* of General Physics and Particular Physics, both of which are inclusive of a large number of sciences. Under "Goodness (Will)" are arranged the sciences "which relate to being as fitted to satisfy the wants of the spirit." These are Eudemonology, Moral Philosophy, and Jurisprudence, with its various subordinate and sub- Peccenini.

sidiary disciplines. Under "Beauty (Æsthetic sentiment)" are classed the sciences "which refer to being as capable of gratifying the spirit and senses." These are Æsthetics, which treats of abstract beauty, and a number of sciences which deal with concrete beauty as exemplified in forms, motions, sounds, and words. Such is the classification of Signor Peccenini. I believe that neither its metaphysical nor its psychological principle will stand examination. Placing the physical sciences after the theological and psychological sciences is in various respects obviously unnatural. Not one of the larger groups seems accurately divided and distributed.

Di Giovanni.

In 1863 appeared also the first edition of Prof. Di Giovanni's *Principii di Filosofia Prima* (the 2nd ed. is of 1878), in which (vol. i. Lez. 3) the sciences are classified as belonging either to Primary or Secondary Philosophy. The former is represented as comprehending Logic, Ontology, Theology, Cosmology, Psychology, Noology, and Ethics; the latter as containing Æsthetics, Philosophy of Systems, Social Philosophy, and Philosophy of History. The learned author endeavoured to show that his classification can be connected with, and conformed to, the ideal formula of Gioberti.

Herbert Spencer's essay on *The Genesis of Science*, originally published in 1854, was largely devoted to the refutation of Comte's views regarding the rational arrangement of the sciences. His own views as to their correlation were expounded in a subsequent essay on *The Classification of the Sciences*, originally published in 1864; and obviously opposition to Comte must have been a considerable motive and factor in their formation. He held that "the sciences as arranged in the succession specified by M. Comte do not logically conform to the natural and invariable hierarchy of phenomena"; that "there is no serial order whatever in which they can be placed, which represents either their logical dependence or the dependence of phenomena"; and that "the historical development of the sciences has not taken place in any serial order." At the same time, he thought that the sciences may be distributed into classes, and endeavoured to show how that may be done on what he regarded as the only true principle of classification—namely, that in each class of colligated facts more numerous and radical characteristics must be included than any of its facts have in common with objects excluded from the class. Now, having regard to this principle, the broadest natural division of the sciences is, he affirmed, that between sciences which deal with the abstract rela-

Herbert Spencer.

tions under which phenomena are presented to us, and those which deal with the phenomena themselves—between sciences which deal with the mere blank forms of existence, and those which deal with real existences. The former class contains Logic and Mathematics, and these are pre-eminently the Abstract Sciences. The latter class is composed of two great groups of sciences, the Abstract Concrete Sciences and the Concrete Sciences. The Abstract Concrete Sciences treat of realities in their elements, or of the real relations implicated in certain classes of facts. Such are Mechanics, Physics, and Chemistry. The Concrete Sciences deal with realities in their totalities, or, in other words, with aggregates of phenomena. They comprehend Astronomy, Geology, Biology, Psychology, and Sociology. "From the beginning, the abstract sciences, the abstract concrete sciences, and the concrete sciences have progressed together, the first solving problems which the second and third presented, and growing only by the solution of the problems; and the second similarly growing by joining the first in solving the problems of the third. All along there has been a continuous action and reaction between the three great classes of sciences."

The classification of Mr Spencer has been criticised by Bain in his *Deductive Logic*, by Renouvier in his *Psychologie*, by Siciliani in his *Rinnovamento della Filosofia positiva in Italia,* and others. It has been adopted with some modifications by Mr Fiske in his *Cosmic Philosophy.* It has been defended against the objections of Dr Bain by Mr Spencer himself in the third edition of his essay (1871). My own criticism of it must necessarily be much briefer than I could wish.

Mr Spencer was probably right in holding that any merely serial arrangement of the sciences must be an inadequate and erroneons expression of their relations to one another. But he can hardly have been correct in supposing that there is no natural series of the sciences at all—none representative either of logical dependence or dependence of phenomena. In fact, he himself recognised a truth which plainly implied that sciences may be arranged in series according to their logical dependence. Mark the following words :—

The three groups of Sciences may be briefly defined as laws of the *forms,* laws of the *factors,* laws of the *products.* And when thus defined, it becomes manifest that the groups are so radically unlike in their natures that there can be no transitions between them ; and that any Science belonging to one of the groups must be quite incongruous with the Sciences belonging to either of the other groups, if transferred. How fundamental are the differences between them will be further seen on considering their functions. The first, or abstract group, is *instrumental* with respect to both the others ; and the *second,* or abstract-concrete group, is *instru-*

mental with respect to the third or concrete group. An endeavour to invert these functions will at once show how essential is the difference of character. The second and third groups supply subject-matter to the first, and the third supplies subject-matter to the second; but none of the truths which constitute the third group are of any use as solvents of the problems presented by the second group; and none of the truths which the second group formulates can act as solvents of problems contained in the first group.

In that passage we are told that the abstract sciences, Logic and Mathematics, are instrumental to the abstract-concrete sciences, Mechanics, Physics, and Chemistry, and that all these sciences of both classes are instrumental not only to such concrete sciences as deal only with mathematical, mechanical, physical, and chemical properties—*e.g.*, Astronomy and Geology, but also to those which are conversant with distinctly new peculiarities—*e.g.*, Biology and Psychology. But if so, on what ground could Mr Spencer maintain that the sciences of Logic, Mathematics, Mechanics, Physics, Chemistry, Biology, and Psychology do not form a logically dependent series? Is Logic not as instrumental to Mathematics as Mathematics to Mechanics or Physics? Is Physics not as instrumental to Chemistry as Chemistry to Biology? How could Mr Spencer contend that Biology is not instrumental to Psychology, seeing that he represented both as sciences of the same class? Astronomy and Geology, however certain and soundly constituted sciences they may be, cannot possibly be ranked among sciences which deal with elements or properties not resolved, or proved to be resolvable, into properties with which more general sciences are occupied. But these two sciences being removed from where they have plainly no right to be, Mr Spencer would seem to have himself constructed a series of sciences of the very kind which, in opposition to Comte, he declared to be impossible. Comte meant no more by calling one science logically dependent on another than that the one placed first is instrumental as regards the one placed last, while the latter is not instrumental as regards the former. If there be a number of sciences dealing with fundamentally distinct phenomena, and so related that every antecedent is instrumental as regards every consequent, and no consequent is instrumental as regards any antecedent, a series of sciences is constituted which represents the logical dependence of its members. Mr Spencer started with denying that there was any such series, but ended by implicitly showing that there was one. His own classification, taken in connection with the passage quoted, was a decisive refutation of what was extreme in his own criticism of the Comtist scheme. So far from having succeeded in overthrowing that scheme, he only at the utmost succeeded in slightly modifying it. There is a logical dependence of the sciences. And why? Just because there is a natural dependence of phenomena. The quantitative relations with which mathematics deals are more general than the mechanical laws which physics brings to light; there can be no chemical combinations unconditioned by physical properties; vital functions never appear apart from chemical processes; and there must be life before there can be consciousness. That remarkable hierarchy of phenomena is a fact which a cloud of abstract language or a covering of subtle reasoning may to some extent and for a short while conceal from our view, but which no language or reasoning can efface or even long obscure. And there being such a hierarchy of phenomena, it is scarcely conceivable that there should be no corresponding hierarchy of sciences.

The terminology of the Spencerian classification has little to recommend it. There is no science which deals with concrete things to the exclusion of abstract relations or with abstract relations to the exclusion of concrete things. All science deals with relations, and is more or less abstract. The concreteness of the objects of the so-called concrete sciences is a concretion of elements and laws which are abstract; and the essential function of these sciences is to discover the abstract factors and operations explanatory of the concrete wholes. Triangles, squares, and circles are as much concrete things with respect to space and its relations as the earth is a concrete thing with respect to matter and its physical properties and laws. The only distinction among the sciences as to abstractness is one of more or less; the only difference one of degree and not of kind. It should be obvious, from the very nature of abstraction, that the word abstract is so entirely a term of degree and relation that it cannot be properly employed to denote distinctions deemed ultimate or specific. But Mr Spencer's use of it was not merely inappropriate; it was misleading, inasmuch as it tended to conceal from view that the chief requirement in a philosophical classification of the sciences is to determine which are simple and fundamental, and which compound and derivative. Comte clearly saw the importance of that requirement; Spencer, unfortunately, did not see it, and so threw together into his third group sciences which are really separated by the deepest and widest of scientific distinctions.

Mr Spencer's reasons for affirming that the so-called abstract sciences, Logic and Mathematics, are more widely separated from all others than any other sciences are from one another, are far from convincing. One is that these abstract sciences deal with relations apart from realities, whereas other sciences deal with realities, and "relations of whatever orders are nearer akin to one another than they are to any objects, and objects of whatever orders are nearer akin to one another than they are to any relations." That Mr Spencer supposed to be self-evident. It is not so. Moral relations are surely much more akin to moral actions than to mathematical relations. If not, there should be a science of moral relations parted by a wide chasm from a science of moral actions. Were Mr Spencer's view correct, the division among the sciences into sciences of relations and sciences of objects should be drawn through the whole scheme of science, instead of being merely made use of to separate two sciences from the rest. In fact, it is quite incorrect, and no division of the sciences ought to be founded upon it. There is no science without both objects and relations. There are no relations without objects. The conception of relations without objects is not an abstract, but an absurd conception. What proof did Mr Spencer produce that the abstract sciences deal exclusively with relations? None at all; he merely said that they deal exclusively with space and time, and that "space is the abstract of all relations of coexistence, and time the abstract of all relations of sequence." But how can there be any relations of coexistence without space, or of relations of sequence without time? Every experience and conception of coexistence presupposes the intuition of space, and of sequence that of time. To call the necessary *conditions* of thought and experience *abstracts* from either is a serious abuse of language. Mr Spencer had, however, another reason for regarding his first division of the sciences as the broadest which can be drawn. The abstract sciences, he said, treat of "the forms in which phenomena are known to us," "the empty forms of things," whereas other sciences treat of "the phenomena themselves," "things themselves"; and "the distinction between the empty forms of things and the things themselves is a distinction which cannot be exceeded in degree." Things, things themselves, are, then, phenomena, phenomena themselves—not noumena, or things in themselves. One is glad to know that, for the word "thing" is by itself very vague and nebulous; but knowing it, one must wish to know also what Mr Spencer can mean by contrasting space and time with things or phenomena. Are these "forms of things" not themselves "things"? Are these "forms of phenomena" not themselves "phenomena"? Yes or no? If yes, why oppose forms and things, forms and phenomena? If no, then there are sciences of what are not things, of what are not phenomena—sciences either of nothings or noumena—and, in that case, Mr Spencer's whole philosophy was a vanity, inasmuch as it was based on the supposition that science is limited to the phenomenal. Further, one may reasonably wish to know in what relevant sense Mr Spencer could call space and time "empty forms." If they are empty, how do the sciences which deal with them bring so much out of them? *Ex nihilo nihil fit.* It is manifestly just because space and time are not empty of quantitative properties and relations that there are mathematical sciences; and manifestly just because they are thus not empty, but contain so many of the fundamental attributes of matter, that the sway of mathematical science is spreading over the whole physical universe, and that physical science tends constantly to become more and more mathematical.

I might proceed to show by a direct consideration of the abstract and abstract-concrete sciences of Mr Spencer that the distance between them is by no means so broad as he affirmed, but that has been already so successfully accomplished by other critics as to be now unnecessary. The abstract sciences, according to Mr Spencer, were Logic and Mathematics; and the former treated of qualitative, the latter of quantitative relations. That Logic treats of qualitative relations was, however, a proposition which he not only failed to substantiate, but failed to make intelligible. He regarded it as conversant in some sense, like Mathematics, with time and space. What, then, are the qualitative, non-quantitative relations either of time or space? Logic is not limited in the way Mr Spencer supposed. Were it unable to deal with quantitative relations there could be no Mathematics. There is even no perfectly accurate Logic which is not quantitative. Logic if simply qualitative may be conclusive, but cannot be absolutely exact.

Mr Spencer's distribution of the abstract-concrete sciences into Mechanics, Physics, Chemistry, and Sciences of Light, Heat, Electricity, and Magnetism seems inferior to Comte's classification of the fundamental sciences, if Astronomy be excluded. Biology and Psychology, if not Sociology, ought to find their places in this group and not among the merely concrete sciences, as although they have concrete applications, they are in their own natures decidedly abstract. There are important differences between Mechanics and Physics, or rather between Molar and Molecular Mechanics; but it is very doubtful if we ought to regard them as two distinct kinds of Mechanics, or two fundamentally distinct sciences. To do so appears an error akin to Comte's separation of Celestial from Terrestrial Mechanics. How objectionable the designation "abstract-concrete" is may be most readily seen, perhaps, in the case of Mechanics, which in itself is as abstract as Geometry, and in its applications is not more concrete.

The distinction between the so-called abstract-concrete and concrete sciences implies a real distinction, but does not coincide with it. The division which should have been drawn is that between fundamental or simple and derivative or complex sciences. If, instead of Biology and Psychology, Mr Spencer had inserted Botany and Zoology into his third group, he would have conformed much better to his own description of concrete science, and would have ranked along with Astronomy and Geology sciences which resemble them much more in scope, method, and general character.

Like Comte, Mr Spencer failed to recognise how broad is the division between physical and psychical science; like Comte also, he assigned no place in the system of knowledge either to Metaphysics or Theology. These peculiarities of opinion followed naturally from his principles, but must, of course, appear serious defects to those whose principles are different.

The discourse of Prof. Zeller, *Ueber die Aufgabe der Philosophie und ihre Stellung zu den übrigen Wissenschaften,* held at Heidelberg on 23rd November 1868,[1] touches thoughtfully on our problem at various points, but does not directly treat of it. The important work of the late Prof. Harms, *Philosophische Einleitung in die Encyklopaedie der Physik,* which forms the first volume of Karsten's *Allgemeine Encyclopaedie der Physik,* and was published in 1869, does not classify or distribute the non-physical sciences. **Zeller.**

VI. FROM BAIN TO WUNDT.

The late Dr Alex. Bain of Aberdeen in the first division of his *Logic* — the volume devoted to *Deduction,* and published in 1870—has dealt with the classification of the sciences with characteristic ability. He started with the affirmation that Science is the perfect form of knowledge, and thus indicated its peculiarities: "It employs special means and appliances to render knowledge true, is knowledge made as *general* as possible, embraces a *distinct department* of the world, or groups **Bain.**

[1] Republished in his *Vorträge u. Abhandlungen.* Zweite Sammlung, 1877.

together facts and generalities that are of a kindred sort, and has a certain *order* or arrangement of topics, suitable to its ends, in gathering, in verifying, and in communicating knowledge." Then, accepting as primary and fundamental Comte's division of the Sciences into Abstract and Concrete, he described the former as the truly fundamental sciences, and as bound to precede the latter. Logic, Mathematics, Mechanics or Mechanical Physics, Molecular Physics, Chemistry, Biology, and Psychology, are what he held to be the fundamental sciences. "In every one of these," he has said, "there is a distinct department of phenomena; taken together they comprehend all known phenomena; and the order indicated is the order from simple to complex, and from independent to dependent, marking the order of study and evolution;" and, further, that, taken collectively, "they contain the laws of every known process in the world, whether of matter or of mind; and set forth these laws in the order suitable for studying and comprehending them to the greatest possible advantage. No phenomenon can be strange to any one thoroughly conversant with these subjects." In Appendix A he has treated very briefly the classifications of Bacon, D'Alembert, Neil Arnott, but very carefully that of Herbert Spencer. The first five, indeed, are disposed of in a few lines.

The criticism of Spencer's scheme seems to me to be a quite adequate and very conclusive study. So far as I am aware, it has not been answered.

Part II.—the vol. on Induction—treats in Book V. of the 'Logic of the Sciences' (pp. 193-367). Here Logic seems to be represented as so absorbed in the other fundamental sciences as not to be itself a science, or the first of the sciences, or in possession of a specific method of its own no less than other sciences. Epistemology, on the other hand, appears to be left out of account. But how can that be justified? Apparently Logic should be preceded by and included in Epistemology rather than the latter should be absorbed in the former. Dr Bain has dwelt instructively on the notions, propositions, definitions, and axioms of Mathematics. His divisions of Mathematics are—(1) Arithmetic, (2) Algebra, (3) Geometry, (4) Algebraic Geometry, which furnishes rules for the embodiment and interpretation of formulæ, and (5) the Higher Calculus, which deals with incommensurable quantities. Mathematics is followed by Physics, and Physics is divided into the Physics of Masses (Molar Physics) and the Physics of Molecules (Molecular Physics). Molar Physics is represented as having Abstract and Concrete Branches. The Abstract Branches comprise — Mathematics of *Motion* (*Kinematics*); Forces *in equilibrio* (Statics), and Forces *causing motion* (Dynamics). The Concrete Branches are—*Mechanic Powers* and *Solid Machinery*, *Hydrostatics* and *Hydro-dynamics*, *Aerostatics* and *Pneumatics*, *Acoustics* and *Astronomy* (pp. 222-233). Molecular Physics assumes masses of matter to be composed of atoms or molecules that attract or repel each other in various modes, and in consequence of which its chief subjects are *Attractions* (Cohesion, &c.), *Heat*, *Light*, and *Electricity* (pp. 233-243). Chemistry follows directly on Physics, and is intimately related to all the departments of Molecular Physics. It is divided into Inorganic and Organic (pp. 242-257). Biology is placed immediately after Chemistry, and defined as the science of living bodies, all such bodies being constituted from elements common to them all. Under that head the structure, functions, various distinctive notions, methods, and hypotheses of Biology are treated of (pp. 258-275). Psychology is represented as the last of the Abstract Sciences; as comprehensive of both animal and human mind; and so intimately related alike to body and mind that they are always concomitant, and every fact of mind has two sides, a mental and a physical. The Science of Character is presupposed by and conjoined with that of Mind. The account of Psychology (pp. 275-286) is throughout remarkably clear and instructive, and so likewise, although more briefly, what is said of the Science of Character (pp. 286-290).

Besides the Fundamental or Abstract Sciences there are also in the scheme of Dr Bain Dependent and Concrete Sciences. There are further distinctively 'sciences of classification,' which include not only Mineralogy, Botany, and Zoology (see Part II., pp. 292-314), but also Meteorology, Geography, Sociology, and Philology (Part I., p. 28). Dr Bain seems to have forgotten, when occupied with Part II., what he had written in Part I. In self-consistency his list of Concrete Sciences should have included seven sciences,—the four in Part I. as well as the three in Part II. Of all the Concrete Sciences he maintained that "no one of them involves any operation but what is expounded in the fundamental or departmental sciences."

Finally, Bain has included in his scheme Practical Sciences. These form not only a large but a most heterogeneous group, including arts like Building and Dyeing, disciplines like Jurisprudence and Political Economy, and sciences which may fairly be held to be themselves fundamental and departmental, as, for example, Economics, Ethics, and Æsthetics. That group is no natural class but an artificial and heterogeneous conglomeration, to which may be added all sorts of occupations, as, *e.g.*, Baking, Brewing, and the like. Like Comte and Spencer, Dr Bain has acknowledged neither Metaphysics nor Theology to be sciences of any kind. For that view I have never found good reasons given. The opinion has been always rested mainly on misconceptions as to what Metaphysics and Theology are, and also as to what should be understood by the terms knowledge, science, and philosophy. Leaving out of account Dr Bain's unsatisfactory conception as to what should be called 'Practical Sciences,' his classification of the sciences properly so called may well be regarded as an improvement on Comte's and much superior to Spencer's.

Cantoni. Prof. Carlo Cantoni, well known by his remarkable studies on Vico and Kant, and the most eminent representative of Neo-Kantian criticists, also sketched a classification of the sciences in his *Corso elementare di Filosofia*,—a work published in the same year as Bain's, and which has gone through at least ten editions. He would divide the sciences, according to the nature of the cognitions which constitute them, into two classes—the ideal or rational and the experimental. And he would further divide them according to their matter or objects into three classes—namely, 1. Those which treat of the fundamental principle and universal conditions of existence—Ontology, Natural

Theology, Cosmology; 2. Those which treat of material things and conditions — Physics, Chemistry, Natural History, Mathematics; and 3. Those which treat of the powers, laws, and actions of spiritual beings, *i.e.*, men—Psychology, Logic, Ethic, Æsthetic, Philosophy of Law, Philosophy of History, and Pædagogy. Whatever merits this scheme may have it may also be held to be defective in that it does not recognise the necessary conjunction of the ideal and experimental in cognition, nor the unnaturalness of placing first the sciences which are most remote and abstruse, nor the error of treating fundamental and derivative sciences as of co-ordinate rank.

Valdarnini. The first edition of Prof. Valdarnini's *Principio Intendimento e Storia della Classificazione delle umane conoscenze secondo Franceso Bacone* also appeared in 1870. The second edition is of 1880. It contains a skilful exposition and energetic defence of the Baconian classification, and gives a brief but meritorious account of a number of other classifications of the sciences.

Peyretti. G. B. Peyretti, who has drawn his philosophy largely from Rosmini, discourses of the evolution and distribution of the sciences in his *Istituzioni di Filosofia teoretica*, published at Turin in 1874. The fundamental division of his classification is into *rational* or *human* sciences, which are conversant with the *intelligible;* and *revealed* or *divine* sciences, conversant with the *superintelligible.* Each of these orders is divided into a *scienza prima* and *scienze seconde.* The primary science supplies to the secondary sciences appropriate principles, and may be regarded as the organising and organic whole of which the secondary sciences are the members. The primary rational science is Philosophy, which is either Theoretical (inclusive of Ideology and Metaphysics) or Practical. The secondary rational sciences are Mathematics, Physics, Chemistry, Mineralogy, Botany, Zoology, Medicine, Jurisprudence, &c. Theology is said to be the primary revealed science, and Dogmatic Theology and Moral Theology the secondary revealed sciences.

The separation of the intelligible and superintelligible, of philosophy and theology, of rational and revealed sciences, as presented in that scheme, implies a very perplexing dualism which Peyretti attempts to transcend by the supposition of "a science of the whole, both intelligible and superintelligible—a synthesis of the sciences"—Encyclopædia. But must not such Encyclopædia be deemed the only true *scienza prima,* and Philosophy and Theology only *scienze seconde?* Besides, how is the synthesis to be effected? Is it by reason or revelation? In either case reason and revelation must stand to each other in another relation than is implied in the contrast which demands a synthesis. The classification of Peyretti unfortunately rests on conceptions of the relation of reason to truth and science, and of nature to revelation, which must render it unacceptable to all but a small class of religionists, and are too likely to lead others to undervalue the really judicious observations which he has made on the forms of knowledge and the stages of its development.

Baldassare Labanca, Professor of the Science of **Labanca.** Religions in the University of Rome, has written many most interesting philosophical works and a still greater number which deal with religious questions. Any student of theology would find it well worth the trouble of acquiring a knowledge of Italian, were it only that he might be able to read the works of Labanca. Of course, his classification of the sciences is all that here concerns us. He advocates what he calls an *inclusive* system of philosophy, in opposition to *exclusive* systems, devotes a chapter of his *Dialettica* (vol. ii. lib. iv. c. i.), published in 1875, to a consideration of the proper encyclopædical arrangement of the sciences. In his view, a logical distribution of *truth* must be the basis of a logical distribution of *the sciences*, seeing that truth is the end of all the sciences. But that is threefold. All truth comes from the ideal world, the real world, or the social world, and is apprehended through reason, sensibility, or testimony. Hence three classes of sciences —the *speculative*, *experimental*, and *documental.*

To the speculative class belong the metaphysical, mathematical, ethical, juridical, political, and æsthetical sciences; to the experimental class, all the sciences called positive—physics, mechanics, chemistry, geology, &c.; to the documental class, the historical, linguistic, geographical, statistical, and economical sciences. All sciences, however, assume certain principles and primary data, and so presuppose and depend on Philosophy. The divisions of philosophy correspond to those of science; hence, *a philosophy of spirit,* comprehensive of the ideal or speculative sciences; *a philosophy of nature,* regulative of the positive or experimental sciences; and *a philosophy of history,* which dominates the documental or social sciences. These three great branches of philosophy spring from *a primary and universal philosophy,* the one root and common stem of the tree of knowledge.

Is that scheme as true and solid as it is neat and symmetrical? No; and for a reason fully acknowledged by Signor Labanca himself. He tells us that he bases his fundamental division merely on the predominance of the traits mentioned, not on the exclusion of others; that the speculative sciences cannot dispense with experience and authority, or the positive sciences with reason and authority, or the social sciences with reason and experience; and that all the sciences are, in fact, mixed, being drawn more or less from all the worlds of truth through all the channels of knowledge; but he contends that the division, instead of being in consequence discredited, is only thereby proved to be in conformity with the inclusive nature of dialectics. Surely it proves rather that a dialectic thus inclusive is incompetent to draw specific distinctions. It would, besides, be difficult, if not impossible, to make out, as regards the particular sciences, even the predominance or preponderance asserted. Other objections suggest themselves, but may be withheld.

The work of Prof. Conti, *Il Vero nell' Ordine* **Conti.** (2 vols., 1876), is very largely occupied with the doctrine of the sciences. The encyclopædic problem is the theme of the eleventh chapter. Science, history, and art are represented as the departments of human knowledge. Science is the first in the order of reflection, but the last in the order of formation. It is to be divided into Philosophy, Mathematics, Physics, and Positive Theology. Philosophy is either *speculative* or *practical*, in the former case including Ontology, Rational Theology, Cosmology, and Psychology; in the latter Logic, Æsthetics, and Ethics. Mathematics is either *pure* or *applied.* Physics comprehends Physics in the special sense of the term, Chemistry, Physiology, and Pathology, and Physical Anthropology. Positive Theology is founded upon authority, and therefore to be entirely separated from the theology which, being based on reason, is a part of philosophy. I leave it to the reader to criticise that scheme for himself.

B. Erdmann.

In 1877 an article of Benno Erdmann on the "Gliederung der Wissenschaften" appeared in the *Vierteljahrschrift für wissenschaftliche Philosophie*, Bd. ii., Hft. i. It is marked by the clearness and penetration characteristic of its author, and although in its general conclusions there may be little that is remarkable, the observations which it contains on the nature and limits of various particular sciences are undoubtedly most worthy of consideration. The sciences as a whole are conceived of by Erdmann as a system conversant with a complex of regular series of elementary data. Each series is represented by a special discipline, and there are as many groups of sciences as there are different kinds of series. The mathematical sciences constitute the first great group, as their series are resolvable into absolutely like elements and purely logical relations. The sciences concerned with causal connection and real evolution are, then, divided into *formal*—those which seek general laws—and *material* or *historical*—those which deal with the processes of change which arise from the interaction of general laws. It is next argued that in the present state of our knowledge we must also distribute them into *mechanical* and *psychical*—*Naturwissenschaften* and *Geisteswissenschaften*—but with the admission that this distinction may eventually be discovered to be unwarranted. After a few general remarks on the formal mechanical sciences, the historico-mechanical sciences—Astronomy, Geology, Anorganology, Organology, Anthropology—are more fully characterised. The sciences held to belong at once to the formal and the psychical class are Psychology and the normative sciences of knowing (Logic and Theory of Cognition), of willing in conduct toward things (Ethic), and of feeling in the appreciation of things (Æsthetic). While Psychology treats of psychical processes as they are, the other psychical sciences just mentioned discuss their validity. The historico-psychical sciences are unfortunately not described and distributed. The sciences even when combined are, according to Erdmann, incomplete; between them and within them there are blanks or gaps which can only be filled up in a hypothetical manner; and there is a discipline — not entitled to be called a science — which has this function—namely, Metaphysics. Besides may be appended Pædagogy and Theology, the former an art based on Psychology and Ethics, and the latter one which undertakes to satisfy the interest of the general understanding in the ultimate questions of knowledge in a way conducive to culture and progress.

These are the findings of Dr Benno Erdmann. Some of them are, I think, not in the least made out. A little reflection on the distinctive nature of Theology, on the character of its relation to the sciences, and on the number of disciplines, some of which are plainly theoretical, which it embraces, should suffice to show that it cannot properly be ranked along with Pædagogy, and regarded as merely a practical appendage to psychical research and metaphysical conjecture. The account given of the function of Metaphysics is more amusing than edifying. If true, she who was erewhile held to be the queen of the sciences is, in reality, but a degraded and untrustworthy handmaiden who mends their tattered garments by patching them with cobwebs. It is obviously, however, not true, for the whole representation given of Metaphysics is but a mutilated and caricatured reflection of the idea of a doctrine of the sciences—

a doctrine which has for aim to trace the limits, note the defects, and exhibit the relations of the sciences, as much without hypothesis or conjecture as possible. In regard to the so-called normative psychical sciences due weight is not assigned to the fact that the validity of the distinctions between truth and error, right and wrong, beauty and deformity, can no more be shown to result from mental than from mechanical processes, and must be the object of investigations of a kind commonly called metaphysical.

Prof. Simone Corleo has treated of the doctrine of the sciences, or, as he calls it, *Sophology*, in his *Sistema della Filosofia Universale* (Rome, 1880). He distributes the sciences into *physical*, *metaphysical*, and *moral*, and gives under each head an ample enumeration of particular disciplines; but he does not show how his classes are related, or group their constituent members, or arrange these members in their natural order of sequence, contiguity, or dependence. The classification is the conclusion of his work. It is preceded by a special treatment of psychology, anthropology, and sociology. The treatise as a whole is a very acute and ingenious exposition of a philosophy of identity. The author has earnestly and skilfully combated atheism, pantheism, and other inadequate representations of the Divine. His name has an honourable place

Corleo.

among Italian patriots in a great crisis of Italian history.

Bourdeau. M. L. Bourdeau, in his elaborate *Théorie des sciences* (2 vols., 1882), resumed the work of Comte in the spirit of Comte, seeking to expound an "integral" or universal science into which shall enter no metaphysical or theological conception. His treatise is one of very great importance, to which, were the publication of my studies on the *scientia scientiarum* continued, I should have frequently to refer. At present, however, I need only state that, like Comte, he arranges what he regards as fundamental sciences in a single linear series; and that series runs as follows: 1. *Positive Ontology* or Logic, the science of realities, employing the method of intuition; 2. *Metrology* or Mathematics, the science of magnitudes, employing the method of deduction; 3. *Theseology* or Dynamics, the science of positions, employing the method of observation; 4. *Poiology* or Physics, the science of modalities, employing the method of experimentation; 5. *Craseology* or Chemistry, the science of combinations, employing the method of integration; 6. *Morphology*, the science of forms, employing the method of comparison; and 7. *Praxeology*, the science of functions, employing the method of connection. The Ontology of M. Bourdeau is mainly a kind of Psychology, and not entitled, it seems to me, to the place of priority which he assigns it. At least one whole department of his Theseology—that of Kinetics—properly belongs to Mathematics. The way in which he distinguishes Morphology and Praxeology, and divides and distributes both, is the most original and ingenious part of his scheme, and I regret that I cannot give it the consideration which it merits. I think it could be shown that the separation of forms and functions, necessary and important although it be within certain limits, is not so radical and far-reaching as he would make it. The new designations which he gives to the methods of the sciences seem as little to be commended as the new names which he applies to the sciences themselves. Of course, the objections which hold good against positivism in general must hold good against the positivism of M. Bourdeau.

The Order of the Sciences, an Essay on the Philosophical Classification and Organisation of Human Knowledge, published in 1882 by Prof. Charles W. Shields, of Princeton, may fairly be ranked among the best of the smaller treatises which have appeared on the subject of which it treats. Its exhibition of the scheme of scientific distribution adopted is clear and skilful; its criticism of other Shields.

classifications is discriminating and incisive. The author successively enunciates and applies to his problem the following principles: "1. A philosophical scheme of the sciences should be based upon the facts which support them, rather than upon the ideas which they involve; 2. Such a scheme should fully reflect all the distinct classes of facts which have been scientifically ascertained; 3. It should exhibit all classes of facts in their actual connections as coexistent in space and successive in time; 4. It should embrace both the empirical and metaphysical divisions of the sciences in logical correlation; and 5. It should have its completion in a general science of all the other sciences, based upon their historical and logical evolution."

A strict application of the first of these principles, he thinks, "would exclude the abstract sciences of Logic and Mathematics from a philosophical classification, and retain them as disciplinal studies, until, by being employed in empirical investigations, they acquire a content of positive knowledge, when they simply become parts and processes of other more real sciences." As regards the second principle, he holds "that the progress of science has brought into view six distinct classes of facts, affording ground for as many corresponding groups of fundamental sciences

—the Physical, the Chemical, the Organical, the Psychical, the Social, and the Religious." In connection with his third principle he maintains that, "although the different classes of facts are distinct and separate, yet they are found succeeding one another in a fixed order of mutual dependence and increasing multiformity, each involving its predecessor, and becoming a condition precedent to its successor; and with such actual procession of phenomena must correspond the normal procession of the sciences." He also lays down a series of what he calls Principal Sciences—Astronomy, Geology, Anthropology, Psychology, Sociology, and Theology — "each Principal Science representing, in a concrete form, the parallel group of Fundamental Sciences to which it corresponds, and including, as its special domain, all of those Fundamental Sciences from which it is not excluded by its immediate predecessor and successor in the series." All these sciences, he argues, exemplify the fourth principle by being half empirical and half metaphysical. And he concludes by treating of the conditions and nature of that terminal science which, as the fifth proposition affirms, must organise and complete all other sciences.

In the following respects these views of Dr Shields fail to command my assent. The ideas of a science may be its facts, as, for example, in

Mathematics, which is truly science in its most perfect form and no merely disciplinal study. Further, ideas may give to the facts their distinctive character. Only if the idea of God have validity can religious facts be more than simply facts of psychology. But for the idea which underlies it theology would have to be included in mental pathology. Again, moral and æsthetic facts seem as distinct from merely psychical facts as social and religious facts. Then, I cannot concur in the acceptance of Comte's doctrine of a single linear series of sciences. The relationship of the sciences is not truly represented when it is reduced to a simple order of sequence. The conception of a series of Principal Sciences parallel to a series of Fundamental Sciences also appears very questionable. Is it not misleading, for instance, to bring together Astronomy and Theology as Principal Sciences, seeing that Astronomy is merely one of a number of sciences of physical facts, whereas Theology is the science of religious facts? Further, while holding that the sciences involve metaphysical ideas or conditions, I do not deem it correct to maintain that they have each a metaphysical part. To do so ignores the connection of the categories, and is inconsistent with the unity and independence of metaphysics. Finally, while accepting Dr Shields's account of the function of the doctrine of science as thoroughly just, I cannot regard the doctrine itself as the terminal science, but only as the first department of philosophy.

Mr H. M. Stanley, another American writer favourably known by his contributions to philosophy and mental science, has published a paper, well worthy of consideration, *On the Classification of the Sciences*, in *Mind*, No. XXXIV., April 1884. It is necessary to leave unnoticed his remarks on the historical classification of the sciences, as also on the distinction between Static and Dynamic Sciences, and to state only the general result at which he arrives as to a logical classification. He places Mathematics alongside of all other sciences, "not as constitutive, but as concomitant"; and then gives the following series of the sciences, as one which is determined by "the principle of aggregation": 1. Chemistry—the Science of Atom; 2. Molecular Physics—Science of Molecule; 3. Molar Physics—Science of Mass; 4. Biology—Science of Aggregated Cell-Masses; 5. Psychology—Science of Individual Man; 6. Sociology—Science of Human Aggregates; and 7. Theology — Science of God. "The order of aggregation," he says, "plainly is: Atoms into molecules, molecules into masses, cell-masses into plants, animals, and men, and these Stanley.

into societies. Nature is thus a combination of wheels within wheels. This classification presents the general order of the dependence of the sciences. If we wish, for instance, to study in Sociology the family, there will be necessarily presupposed a knowledge of the human individual as a psychical whole; and this presupposes a study of the human animal, and this of the cell, and this of masses, molecules, and atoms. Herein is a 'hierarchy of the sciences.' If this be the order of dependence of the sciences, it must also be the order of their completion, the higher sciences necessarily waiting on the lower. Again, it is also the order of increasing complexity, as has been exemplified throughout. It is also the order of increasing speciality and concreteness, in that it is a logical order of increasing intension and decreasing extension. A number of objects decrease, and numbers of attributes increase. It is also the order of recognised rank."

On this simple yet ingenious scheme of Mr Stanley the following criticisms may be offered: First, it is not shown that Mathematics only is so concomitant with the other sciences that it cannot be simply placed in a series of the sciences. The same is true of Logic, inasmuch as all other sciences are built up by logical processes. The same is true even of Theology, inasmuch as all other sciences furnish materials for Theology. Secondly, the conception given of the nature and position of Chemistry seems untenable. Chemistry is defined as the Science of the Atom, and as such is regarded as the first constitutive science. In reality, Chemistry has not yet proved the existence of the atom. The atom is still only an assumption, and may turn out to be a pseudo-metaphysical fiction. And should its existence be scientifically established, it is most improbable that it will not be found to have properties and relations of a mechanical order, simpler and more general than its chemical characteristics. Chemistry has not to do with atoms more than with molecules and masses. It has to do with the analysis of compounds into elements and the synthesis of elements into compounds. It is, as M. Bourdeau says, the science of combinations. Thirdly, the principle of aggregation is insufficient and unsuited for the classification of the sciences. It is just because there are distinctions of things which cannot be explained by aggregation that there are distinct sciences. If life and mind could be shown to be simply aggregates, Biology and Psychology would be at the same time resolved into Chemistry. Sociology can have no claim to be more than a department of Psychology unless it can be shown to be more than "human aggregation." The idea of God, in which Theology is rooted, is not that of an aggregate.

D. G. Thompson.

Mr Daniel Greenleaf Thompson, also an American, exhibits a tabular scheme of classification of the Sciences in his *System of Psychology* (vol. i. p. 76, 7), published in 1884. His main division is into (*a*) Sciences relating primarily to the extended—Non-Ego Sciences; and (*b*) Sciences relating primarily to the unextended—Ego Sciences. Class A is subdivided into Physics and Biology, each of which is represented as including various Abstract and Concrete Sciences. Class B is subdivided into Theoretical and Practical Sciences. The former are subdivided into: 1. Sciences of Mind in its relations to itself, comprehending the Abstract Sciences of Logic, Mathematics, and Æsthetics, and the Concrete Sciences of Psychology and Ethnology; and 2. Sciences of Mind in its relation to other Minds, comprehending the Science of Human Communication and Sociology, with its related group of studies. The scheme, it may be perceived, is of an external and artificial kind. It rests on no principle, proceeds on no consistent method, and is pervaded by no general philosophical conception. It counts various sciences twice, first as theoretical and next as practical, and it is not apparent why all are not so dealt with, while it seems almost absurd to confine the distinction of Theoretical and Practical to the Ego-Sciences. Mathematics and Logic are placed after all the physical sciences, although both are plainly presupposed by all these sciences. While prominence is given to the questionable distinction of Abstract and Concrete Sciences, the much more significant one of Fundamental and Derivative Sciences is ignored. No room is found for Theology. Several other errors of Bain and Spencer are reproduced.

It is now necessary to give some account of the views of M. E. De Roberty on the subject in hand. He is a native of Russia but lives in Paris, and is a most industrious as well as very able French publicist. He is a thorough positivist, but very far from a mere Comtist or, indeed, a mere disciple of any teacher. He often rejects Comte's conclusions and substitutes for them very different views of his own; and, in fact, is one of the most independent as well as one of the most interesting and instructive contemporary thinkers of the positivist school. Of all criticisms of Comte and contributions to positivism those of Roberty are, perhaps, on the whole, the most thorough and suggestive. **De Roberty.**

His views on the classification of the sciences are to be found chiefly in his *La Sociologie*, 1881. There he has distributed all that he regards as sciences into *four groups*. The reason given for doing so is that the sciences of each of those groups rest on different ways of observation. The sciences of the *first group* are held by him to be the *mathematical sciences*, on the ground that they rest on *simple intuitions* or *self-evident axioms*. *Astronomy*, on the other hand, is made to do duty as representative of a *second kind of science*, or perhaps group of sciences, because based on *pure and simple observation*. *Physics and Chemistry* are adduced as constitutive of a *third class*, and one of special interest inasmuch as dependent not only on observation but on *observation conjoined with experimentation*. And, further, there is a *fourth class*, the sciences of which are designated by Roberty *descriptive sciences*, because grounded on what he calls *scientific description*,—a process on which he has dealt at considerable length and to which he attaches great importance. In that last class he has included *Mechanics*, *Biology*, *Psychology*, and *Sociology*. To Sociology he assigns the same place, and attributes much of the same importance, as Comte had done. The definitive co-ordination of the sciences he holds to be the task to which the *Philosophy of the Sciences* is bound to devote itself,—a task which is still in the future but will not fail to be accomplished. Six years later than *La Sociologie* appeared his *L'Ancienne et la Nouvelle Philosophie* (1887), which was followed by five works, the parts of a single system of thought, and the titles of which are *L'Inconnaissable* (1889),

La Philosophie du Siècle (1891), *Agnosticisme* (1892), *La Recherche de l'unité* (1893), and *Auguste Comte et Herbert Spencer* (1894). They are all meant to be contributions to a true philosophy of the sciences, a *scientia scientiarum*, a whole of positive sciences alone, one on which each positive science depends for its development on the antecedent sciences, and on which all real philosophy depends exclusively on all real positive sciences. They are all meant also to convince their readers that "the whole of religion and the whole of philosophy so-called" have nothing in them of the real nature of science; that there is no such thing as theological or metaphysical science; that even the so-called criticism of Kant, the positivist agnosticism of Comte, the conditioned or relativist agnosticism of Hamilton and Mansel, and the evolutionist agnosticism of Spencer are all forms of pseudo-science or of philosophy falsely so called.

The courage and self-consistency of Roberty in extruding all theology and metaphysics from what he considers knowledge or science, and his perspicacity in showing that very much of what has been affirmed by modern Agnostics is as nonsensical as anything of an analogous kind which can be laid to the charge of medieval scholastics, are worthy of recognition, but he has quite failed to prove all metaphysics and theology to be of an agnostic, unscientific, or anti-scientific character. Rightly understood, both metaphysics and theology may be sciences. The exclusion of all theology and metaphysics, of all religion and philosophy, from the rank and nature of sciences, is a serious defect in a classification of the sciences. The views of Roberty and others to the contrary are somewhat fully dealt with in my *Croall Lectures on Agnosticism for* 1887-88. Roberty's first group of sciences are *the mathematical.* Some of those sciences, however, are among the latest, and, alike on historical and rational grounds, it may be questioned whether any of them were the earliest. Logic, for example, may perhaps have preceded any of them both in India and Greece. It is somewhat difficult to conceive how mathematics could have arisen until preceded by a considerable knowledge of knowledge, a clear apprehension of the axioms on which mathematics rest, and of the rules and processes of reasoning. Scientific knowledge has in almost all departments so grown out of ordinary knowledge that it is difficult to determine where the latter has ended and the former begun. Further, Roberty describes *astronomy* as representative of a second group of sciences on the ground that it is a *science of pure and simple observation.* But is it so? What would have become of astronomy were it confined to observation and left unaided by calculation? Other questions are suggested by Roberty's classification. My readers may raise and answer them for themselves. His scheme is a meagre one compared with many others that I have already noticed. That does not, however, much affect the value of his writings, which I wish were more widely known in Britain.

The name of Wm. Wundt is much more widely known than that of De Roberty. Although born in 1832, Wundt is still an indefatigable teacher and experimentalist. Physiology has doubtless been the main subject of his studies, seeing that as privat-docent and professor he has publicly taught it for the long period of forty-seven years, but he has also by original investigations left his mark on many of the chief sciences. Even on logic, ethics, and psychology he has written most elaborate and very valuable treatises. It is only natural, therefore, that he should have occupied himself earnestly with the problem of the relations of the sciences to one another. His range of knowledge must be greatly wider and more exact than was that of Comte. If less of a philosopher than was Spencer, he is much more of a scientist. The works in which Wundt.

he has treated of the classification of the sciences are his *Logik*, Bd. ii. (1883), his *Philosophische Studien*, Bd. v. Th. 1 (1886), and his *System der Philosophie* (1889). They show that he has thoroughly realised the importance of classification of the sciences, and of the dependence of the sciences on philosophy.

Perhaps it is in the last of the works mentioned that he has most completely expounded and defended his conception of philosophy as 'a science of all the positive sciences,' as 'the universal science which has to do with the cognitions obtained by the particular sciences into a consistent system.' His *Logic* is described by himself as 'an investigation of the principles of knowledge and of the methods of scientific research.' Hence its first volume is expressly designated an *Erkentnislehre* and the second a *Methodenlehre*,—the former being regarded as the general theory of logic or of real and formal investigation and reasoning, and the latter as a study of the principles, methods, and acquisitions of the special sciences. In the second edition of 1895 the *Methodenlehre* was greatly enlarged and elaborated so as to be much superior to any corresponding chapters in J. S. Mill's *Logic*. The volume consists of four main sections with subdivisions. It begins with 'a general doctrine of method' (pp.

1-73), and then expounds the logic of mathematics, treating first of its method so far as general, and then in the following order of succession the special methods of arithmetic, geometry, functions, and infinitesimals (pp. 74-219). The logic of the natural sciences is similarly dealt with: an exposition of the general foundation of natural investigation being first given, and then in due order an exposition of the special logical methods of physics, chemistry, physiology, and biology. The logic of the mental sciences is dealt with in the same manner. The bases common to them all are first laid bare, and then those of the historical and social sciences are specially described. The volume is brought to a close with an elaborate exposition of the methods of philosophy (pp. 478-620).

As already said, Wundt has also dealt with the classification of the sciences in his *Philosophische Studien*, Bd. v. Th. 1, 1886. There he divides the general system of the sciences into I. Particular Sciences, and II. Philosophy, and subdivides both. I. The Particular Sciences he distributes into two great groups—Formal Sciences and Realistic Sciences. (A) Formal sciences are the mathematical sciences, and of these a detailed enumeration and description are given. (B) Realistic sciences are subdivided into two sections — viz., (*a*) Physical sciences and (*b*) Mental sciences. The latter are subdivided thus :—(*a*) Theory of the phenomena of spirit (*i.e.*, psychology under its different forms) : (*b*) Sciences of the products of spirit (philology and social sciences) ; and (*c*) Science of the development of the products of the spirit (history under its different forms). II. Philosophy itself is thus subdivided :—(*a*) Theory of knowledge (both formal and realistic) ; and (*b*) Theory of principles, which under its general form is metaphysics and under its particular forms is philosophy of nature and philosophy of spirit.

A still later attempt of Wundt's at the distribution of the sciences is to be found in his *System der Philosophie* (1889). The view of it given by Prof. Ladd of Yale in his *Introduction to Philosophy* (1891) is so brief, exact, and accurate, and so likely to be better than any I could myself produce, that I shall venture to avail myself of it.

The most recent important work aiming at a system of philosophy is by Wundt. As might be expected from its author, this treatise on synthetic philosophy is everywhere conceived and executed in a spirit of fidelity to the method and results of the particular sciences. Wundt regards philosophy as a universal science, having for its problem to unite the cognitions of the particular sciences into a consistent system. On account of the relation in which it stands to these sciences, its divisions must be based on the division of the sciences. Two main problems are, therefore, given to philosophy in its efforts to treat synthetically all the particular sciences. The first of these problems relates to knowing in a process of becoming; the second, to knowing already become (*Wissen, Werdende,* and *Gewordene*). Hence the two main divisions of philosophy are (1) Science of Cognition, (2) Science of Principles. These two divisions are then developed into a scheme, which may be tabulated as follows:—

Division of Scientific Philosophy.

- I. Science of knowledge.
  - 1. Formal (Formal Logic).
  - 2. Real.
    - A. History of Knowledge.
    - B. Theory of Knowledge, which in connection with formal logic constitutes Logic in the wider meaning of the word, is then further subdivided into—
      - (*a*) General Theory of knowledge.
      - (*b*) Theory of Special Methods as applied to scientific investigation.
- II. Science of Principles.
  - 1. General, or Metaphy sic.
    - The systematic exposition of the fundamental conceptions, and fundamental laws of all science.
  - 2. Special.
    - A. Philosophy of Nature, which is subdivided into—
      - (*a*) General Cosmology, and (*b*) General Biology.
    - B. Philosophy of Spirit, which has three subdivisions—
      - (*a*) Ethics, (*b*) Æsthetics, and (*c*) Philosophy of Religion.

On the foundation of the three divisions of the Philosophy of Spirit, and with the help of a comprehensive survey of human development, stands the Philosophy of History. Its aim is to give a picture of the whole external and internal life of man.[1]

Wundt's classification of the sciences merits, I have no doubt, a fuller exposition of it than has just

[1] Ladd, pp. 167, 168.

been given. A criticism of it I shall not undertake, although on several points it seems to be not beyond criticism. The scheme has, I think, a considerable number of defects. Its merits seem far from equal to those of the work done by the author of it on the methodology of the sciences included in it. It is on the latter, not on the former, that Prof. Wundt's labours are of such very exceptional value. Only experts, and experts of an extraordinary range of knowledge, can be expected fully to appreciate how great those merits are. As a general review of Wundt's conclusions as to the classification, logic, and system of the sciences I know none better than Prof. Venn's in *Mind*, vol. ix. pp. 451-468. To it I refer my readers.

VII. FROM MASARYK TO KARL PEARSON.

Masaryk. T. G. Masaryk, professor in the University of Prague, in 1866 published in the Bohemian language a book on "the classification and organisation of the sciences." Fortunately a German translation appeared in the following year. It would well deserve translation also into other European languages, as there is scarcely any other work so likely to serve well as an introduction to as many sciences; for, although its author modestly acknowledges that only on some sociological and psychological departments of research can he make any claim to write as an expert, he has obviously made a thoughtful general study of most of the principal sciences, and acquired an adequate acquaintance with the literature regarding them. The authorities on which he relies are of a good kind. With British philosophical literature he is exceptionally well acquainted. The English authors to whom he refers most frequently are Bacon, Bain, Faraday, Rowan Hamilton, Sir Wm. Hamilton, Hume, Locke, J. S. Mill, Newton, H. Spencer, and Whewell; the French, DesCartes, A. Comte, Pascal, and Roberty; and the German, Du Bois-Reymond, Dilthey, Fechner, Harms, Kant, Leibniz, and Wundt. That Italian authors are so much overlooked is to be regretted.

The German title of Masaryk's treatise is *Versuch einer Concreten Logik*, and his introductory remarks are clear and relevant as to the need of a classification and also an organisation of the sciences. With regard to the character of classification, while affirming its necessity, he allows that there is something artificial in every classification, and that neither evidence, certainty, nor method can be its sufficient principle. The order and relationships of the sciences ought to be determined by the nature of their objects. Theoretical and practical sciences, however, are to be separated. There is the widest distance between them. Abstract and concrete sciences are less so, both classes being theoretical sciences. All these sciences—abstract and concrete alike—are occupied with the natures of certain kinds of objects, the systems of truth that may be elicited from special spheres of knowledge. What the so-called practical sciences aim at is the attainment of desired ends, the accomplishment of purposes deemed useful. All sciences may be applied to several uses, and all arts may be more or less related to some science or sciences. To enclose them in the same scheme cannot be rightly effected, but merely made to seem so, by a cross and confusing division. The study of the sciences is one thing, the application of them to ends and identification of them with arts another. Masaryk's so-called 'practical sciences' seem to have been counted by him as both seven and twelve. There might, I think, have been many more. His list of them is as follows: A. Calculation and Measurement. Descriptive Geometry. Theory of industrial and imitative Arts; B. Technology in widest sense (Rendering serviceable the forces of nature); C. Physical and curative education (Phytotechnic, Zootechnic, Medicine, and Hygiene); D. Training of the character and understanding (Pædagogic and Didactic), Politics, and Ethics (as science of the complete guidance of life); E. Practical Grammar (Mastery of Language); F. Practical Æsthetic; and G. Practical Logic. Such is Masaryk's enumeration of so-called 'practical sciences.'

Obviously some of them would have been better placed among *arts*, while others are as properly *sciences* and should have been so designated. Sciences and arts may be intimately connected, but to call either arts sciences or sciences arts is an error, and must lead to confusion as it has obviously done in Masaryk's scheme. That scheme owes more to Comte than to any one else, and indeed so much that the author of it may be fairly regarded as a Comtist, a very independent and sagacious one however, who cannot be reasonably charged with having taken the views of Comte, or any one else, without close and careful consideration. He has rejected even Comte's linear series of the sciences and substituted for it a binary classification, although his own classification thereby loses the sort of unity which pervades Comte's scheme, and to which more than anything else that scheme has owed its popularity. But for its simplicity Comte's classification would never have been preferred to a considerable number of the more complex schemes that have been already described in our pages.

As already stated, Masaryk distributes his 'theoretical sciences' into 'abstract sciences' and 'concrete sciences.' There has been much controversy as to what should be meant by the terms 'abstract,' 'concrete,' and also 'abstract-concrete.' Comte, Littre, Spencer, and others have been engaged in it without arriving at any very definite or important result. There is no mere abstractness or mere concreteness in the objects of any of the sciences. The term abstract-concrete should imply that and neither more nor less. The division or classification of sciences into abstract and concrete cannot be a complete division, a perfect classification. It may, however, be none the less but all the more instructive on that account, as showing how intimately all sciences are related. Prof. Masaryk attaches great importance to Comte's doctrine of a hierarchy of sciences, — a closely connected series of fundamental sciences. Substantially he adopts it as a whole, yet obviously after a close and independent study of it. Hence he is often accurate where Comte was not, and brings to light what Comte had left in darkness. All the sciences of the hierarchy are, of course, represented by him as abstract sciences, — not concrete and still less so-called practical sciences. Hence it is now necessary to indicate what in his scheme of classification are the abstract sciences and especially what are the sciences of the hierarchy. The following table may suffice :—

The Theoretical Abstract Sciences.

A. The Sciences of the Hierarchy (Fundamental Sciences). The idea of a hierarchy of the sciences was first clearly set forth in the *Pansophiæ Diatyposis* (1645) of Comenius.

I. *Mathematics.*—To it is assigned by Masaryk precedence in the hierarchical sciences and consequently of all other sciences. His description and distribution of the mathematical sciences seem to be about as accurate as could possibly be given in fifteen pages (71-86) by one professedly not a mathematical expert; and show how carefully he has utilised not only the well-known works of Comte, Bain, and Wundt so far as they bear on the subject, but also such works as Baumann's *Lehren von Raum, Zeit und Mathematik in der neuesten Philosophie*, Clifford's *Common Sense of the Exact Sciences*, Cantor's *Vorlesungen über Geschichte der Mathematik*, De Morgan *On the Study and Difficulties of Mathematics*, Duhamel's *Des Méthodes dans les sciences de raisonnement*, Kroman's *Beiträge zu einer Theorie der Mathematik und Physik*, and Schmitz-Dumont's *Die mathematischen Elemente der Erkenntnisstheorie*. Mathematics is, however, a very comprehensive term. It is the name not merely of a science but of a system of sciences, and these closely interrelated sciences, each of which rests on a fundamental idea or ideas, and has a correspondently different method. Space, magnitude, figure, number, time, motion, direction, rate, limit, &c., are all foundations of mathematical reasoning, and all mathematical sciences have so far their own distinctive methods. Arithmetic and Geometry are very different both as to matter and method from the Calculus and Kinematics. That is not sufficiently indicated by Masaryk. He has, however, clearly stated the advantages which the mathematical sciences have in important respects over all other sciences, and also their limitations.

II. *Mechanics.*—According to Masaryk it is the second hierarchical science; one which has very much in common with, and is to a great extent dependent on, Mathematics. It has even been often included among the mathematical sciences. Mach in a treatise on 'the development of mechanics' has contested its right to be so placed, and Masaryk deems his argumentation probably conclusive. Perhaps he is right in thinking so, but certainly Mechanics is both abstract and concrete, both quantitative and qualitative, and cannot be denied to be on the borderland between mathematical and physical science, and to lie almost as much within the territory of the one as of the other.

III. *Physics.*—That third so-called hierarchical science is comprehensive of a large class of sciences, namely, all those which deal with inorganic physical things, or, in other words, with the properties and changes of matter in their molecular constitution, and therefore with hardness, elasticity, cohesion, &c., as also with heat, light, sound, electricity, magnetism, &c. All the sciences referred to are occupied with these objects, their properties, and effects. They are all inductive sciences and dependent on observation and experimentation. Masaryk declines to arrange the departmental physical sciences in any serial order. He regards Comte's attempt to do so as a failure.

IV. *Chemistry.*—It seems strange that Chemistry should not have been included among physical sciences but ranked as an hierarchical science. In its present condition even it seems closely akin to and dependent on the physical sciences, and appears likely to be much more so in the future. What separates Chemistry from Physics as described by Masaryk is that while physical processes leave the material structure of things ordinarily unchanged, chemical processes leave a profound and lasting change. In other words, what is distinctive of Chemistry as compared with Physics is what is called 'chemical affinity,' a peculiar and as yet altogether mysterious ability of matter to enter in its smallest parts into an intimate connection of a kind confined to Chemistry alone. Its products are completely new. No other science apparently takes us so deeply into the nature of matter. A complete knowledge of the evolution of molecules may go far to explain the evolution of worlds. The infinitely little may be a key to acquaintance with the infinitely great. Experimentation has a large place in Chemistry. What *measuring* is in Geometry, *weighing* may not unreasonably be said, as it is by Masaryk, to be in Chemistry.

V. *Biology.*—To this fifth hierarchical science in Masaryk's scheme both Physics and Chemistry are represented by him as subservient, while holding great injury to have been done to it by a crude materialism in unreasonable attempts to explain life and its operations by inadequate causes. A completely satisfactory method of studying it is held to have been as yet far from adequately ascertained. Mere conjectures and conflicting hypotheses abound in it. Its province is an extremely wide one, including not merely a single science but many sciences, as, *e.g.*, Anatomy and Physiology, Botany and Zoology, &c.

VI. *Psychology.*—It is closely connected with and largely dependent on Biology. Life is presupposed in every psychological process. That life has originated out of mere matter (if there be such a thing), has not been fully proved, but no one doubts that where there is no vitality there can be no mental states. Thought, feeling, and volition in every form, all phases and stages of consciousness, presuppose life, not death. Of all the mental sciences Psychology is the fundamental science—the *Grundwissenschaft.* Masaryk's treatment of it (in pp. 116-138) seems to be very judicious.

VII. *Sociology.*—Like other positivists, Masaryk regards Sociology as the crowning hierarchical science, and naturally deals with it at much more length than with any preceding science. He adopts Comte's division of it into Social Statics and Social Dynamics, and also distributes its contents into Theoretical and Practical Sociology. Its connections with, and bearings on, other sciences are likewise traced, and the history as well as probable results of its development and findings are referred to. Biology, Psychology, and Sociology are the inseparable stages in a vast and complex system of evolution.

B. Outside of the hierarchy three other abstract sciences are recognised by Masaryk—namely, VIII. *Philology* (*Sprachforschung*, including *Sprachlehre und Grammatik*); IX. *Æsthetics;* and X. *Logic* (*i.e.*, Abstract Logic). I do not deem it necessary to remark on that part of Masaryk's scheme, nor on his plan or views of the system of the concrete sciences. It must suffice that I enumerate them as given by himself: C. Concrete Sciences. 1. Geometry; 2. Astronomy (Chronology), Acoustics (in part), Hydrostatics, Hydrodynamics, Aërostatics, &c., Cosmography (Astro, Geo, and Oceanography), Cosmology (Astrogeny, Geology, &c., also Cosmical Physics, Chemistry, Astro-physics and Astro-chemistry, Geo-physics and Geo-chemistry, &c.); 3. Botany and Zoology; 4. Concrete Psychology, Ethnology, Political Sciences, Political Economy (including Statistics), and History (both Universal and Special); 5. History of Language; 6. Theory of Arts; and 7. Concrete Logic. All the so-called Concrete Sciences are represented as in one direction or connection closely related to the Abstract Sciences, and in another to the Practical Sciences.

Supplementary to the section of Masaryk's system of the sciences, as above described, are two sections of reflections exclusively on the concrete and practical sciences. Book v. of his work is a statement of his Philosophy understood as equivalent to Metaphysics. Theology he does not admit to be a science or group of sciences. But he treats it respectfully, and acknowledges it to have been a chief condition of scientific progress. He has written a valuable treatise, and discussed in it with varying degrees of lucidity and thoroughness a great number of questions and problems as to the classification and organisation of the sciences. That he has often failed to arrive at definite or accurate conclusions, I am not prepared to deny. To excite thought, however, is often a greater benefit than to satisfy it.

M. Adrien Naville, a worthy son of the illustrious Genevan philosopher, M. Ernest Naville, has earnestly and repeatedly occupied his mind with the subject under consideration. In 1888 he published a *Nouvelle Classification des Sciences;* in 1898 he gave an excellent restatement of *Le principe général de la classification des sciences* in the German philosophical periodical *Archiv für systematische Philosophie*, iv Band, Heft 3, 1898; and in 1901 a second edition of the work which appeared in 1888 is spoken of by the author as "completely recast" (*entièrment refondue*). He describes the purpose of the work so long dealt with as being to trace the boundaries of the special sciences, to distinguish the fundamental notions of each of them, and to mark the relations which connect them. His mode of distributing them has, so far as I am aware, the merit of originality, one now becoming rare among the classifiers of the sciences. It is by grouping the sciences **A. Naville.**

around three questions which he regards as fundamental.

The sciences he maintains are "wholes (*des ensembles*) of answers to questions put by the human mind, and the deepest differences between the sciences are those which result from the answers to the questions laid down." But the fundamental questions referred to are in his opinion just those three: 1. What is it *that is possible?* 2. What is it *that is real?* and 3. What is it *that is good?* Hence he holds that there are three great classes of sciences; and that those sciences which answer the first question are the sciences of *limits* and of the necessary relations of *possibilities*, or, in equivalent terms, the sciences of *laws;* those which answer the second question, the sciences of *possibilities realised*, the sciences of *facts;* and, further, those which satisfy the third question—namely, the sciences of possibilities the realisation of which would be good, or, in equivalent terms, the sciences of *ideal rules of action.* His scheme of classification is entirely dependent on his principle of classification.

His *Tableau* of the former is regulated by the latter, and determines his distribution of the sciences under the three headings—I. Theorematics; II. History; and III. Canonics. As belonging to I. *Theorematics*, he mentions the following sciences:

(1) Nomology, (2) Arithmology, (3) Geometry, (4) Kinematics, (5) Physico-Chemistry, (6) Biology, (7) Psychology, and (8) Sociology. He acknowledges, however, that there may be many more sciences of mere laws, and even an indefinite number of them. What he regards as the science of laws under an absolutely abstract form is what he calls nomology; arithmology (arithmetic and algebra), geometry, and kinematics are at once mathematical sciences and sciences of law; but there are other mathematical sciences, and, even if there were not, there is a vast interval between the mathematical and the physical sciences, and a still vaster between the former and psychology and sociology. That psychology and sociology are occupied merely with the possible, not with the real, is extremely questionable, and indeed M. Naville himself admits that we do not yet possess a truly theorematic psychology or sociology; that they are not universally considered as sciences of laws, but are, on the contrary, largely composed of historical generalisations derived from experience. Herbert Spencer placed them in the same class as astronomy, geology, mineralogy, &c., which are certainly more occupied with the real than with the possible. That the mathematical sciences are sciences of possibilities and theorems and not of realities or facts is not likely to be denied, nor will it be doubted that they are members of a

very distinct and special group of sciences which, although not wholly unrelated to psychology and sociology, are no more related to them than to many other disciplines which M. Naville himself does not include among his so-called theorematic sciences. M. Naville's distinction between laws and facts, possibilities and realities, seems to me to be a real and important one, but also one which he somewhat misapplies and makes too much of.

II. *History*, the second great section of his scheme of classification, is defined by him as the science of realised possibilities or facts. The signification given to it is very comprehensive, and yet, as we have seen, sociology is not included in it but in theorematics, although it surely has as much right to be regarded as an historical discipline as most of those studies which M. Naville has represented as actually included in history. His reason for regarding history as he does is that it is the kind of knowledge or science in which the question, What is that which is real? is solved or in the way of being solved. The real is part of the possible, the possible so far as realised, what presupposes no mere conditions, no contingencies, no ifs. It is concerned only with facts and composed only of categorical affirmations. Further, according to M. Naville, history is not, strictly speaking, a class of definite and separate sciences, but, as he himself says, "a single science without rigidly distinct divisions, because in concrete reality all acts upon all, so that in place of a series of different sciences there are only parts of one science." Hence the parts of history thus understood may be innumerable, and, as they already are in number and character, may be held to constitute the chief objects of human study. Naville's list of them is—(1) Astronomy, (2) Geology, (3) Mineralogy, (4) Botany, (5) Zoology, (6) Anthropology, and (7) Human History, political, moral, judicial, economic, linguistic, literary, artistic, religious, &c. And they are all obviously to a large extent of an historical character. But are they more so than say Sociology or even the History of Mathematical Sciences? Geometry, Biology, Psychology, and Sociology have all histories simply as accounts of them as evolutionary or progressive studies, and their objects would also have had histories had there been no human beings to study them.

III. *Canonics* is the third and last section of Naville's classification of the sciences. He holds it to be a scientific group essentially different from Theorematics and History. It is meant to be the answer to the third great scientific question, which is also the chief practical question, and to include all sciences of the rules of human activity which expressly tend to the realisation of the best possible. A threefold division of it is given. The first (*a*) is *Morale*, the general theory of aims, the system or doctrine of rules relative to the choice of chief ends. Its function, according to Naville, is to study the different aims possible, so as to estimate aright their comparative and complete value;—aims held by him to be of four kinds,—namely, 1°. satisfaction for self, 2°. satisfaction for others, 3°. truth (knowledge) for self, and 4°. truth for others. He leaves it to *la morale* itself to determine the value of all special investigations into the nature of the good, and to show how their findings may be and should be combined. There are, however, in his conception of Canonics two other departments than Morale, a second and third. The second, (*b*) *Theories of the arts*, may be indefinitely numerous, inasmuch as they are held to include all theories which endeavour to formulate rules for selection of the most suitable means to attain ends of every kind; all arts associated with the various species of knowledge or games of chance; logic and didactic; industries, medicine, &c. Finally, as a third division of Canonics there are said to be (*c*) *Moral Sciences;* —sciences said to be composed of rules for the choice of the means best adapted to realise in a harmonious way human ideals. Pædagogy and the Law of Nature (or Reason) are the examples given of them. The former seeks by all attainable means to develop to the utmost and in the most harmonious way whatever elements for good are contained in germ in the natures of children. The latter seeks to ascertain how the State ought so to constrain and regulate the power intrusted to it as to contribute as much as possible to the physical, intellectual, and moral development of all classes in a nation.

M. Naville's classification of the sciences has now been described and as far as possible in his own words. My readers may criticise it for themselves, and decide, say, whether the section of Canonics is satisfactory or the reverse. Before coming, however, to a definitive conclusion even in regard to Canonics, the seemingly weakest part of his scheme, they would do well to take into account that M. Naville published in the *Revue Philosophique* (No. 1, Jan. 1897) a very able essay on *Économique et Morale*, which may be held as a valuable contribution to what would otherwise have rather discredited his whole system, whereas now even Canonics may be deemed not unworthy of consideration.

De la Grasserie.

In 1893 M. Raoul de la Grasserie published his *De la classification objective et subjective des sciences, des lettres, et des arts.* It is an elaborate work of more than three hundred pages, and obviously the result of long and earnest studies. Its author has attempted, however, an almost impossible task by undertaking to deal with three so distinct classes of objects as arts, letters, and sciences, and with two contrary kinds of method, a subjective and objective. Arts and letters are certainly not wholly independent of or unrelated to the sciences, but they are not sciences nor, perhaps, more dependent on the sciences than the sciences are on them. The *subjective method* of De la Grasserie is any suitable order of method for a desirable course of education. His *objective method* is a quite different process. It is a tracing of the order of succession and dependence of the sciences in accordance with their own natures. As I have already so far criticised the classifications of Bacon, D'Alembert, and Ampère, in which arts, letters, and sciences are included, it seems to me unnecessary to dwell on what is akin to them in M. de la Grasserie's scheme. Of course he has not only studied what he knew to have been carefully attempted by the most eminent of his predecessors, but has also sought to appropriate and utilise what seemed to him to have true findings. Those from whom he has derived most are Ampère, Comte, Spencer, and Wundt.

He has accepted as highly important the distinction between sciences of matter and of mind, or what Ampère called cosmological and noölogical sciences. It is a distinction which few thinkers, if any, have either altogether overlooked or rejected. It is not a distinction, however, which can legitimately carry us very far. In proof I refer my readers to my criticism of Ampère's method of bifurcation based on the distinction. See pp. 79-82. Grasserie also adopts what he calls Spencer's 'luminous division of the sciences' into abstract sciences, abstract-concrete sciences, and concrete sciences. (*a*) By Abstract Sciences are meant those sciences which, like Logic and Mathematics, treat of ideals or unoccupied forms of relations in which phenomena are known to us; (*b*) By Abstract - Concrete Sciences those which, like Mechanics, Physics, and Chemistry, treat of real relations or the relations among realities to which different modes of matter and motion conform; and (*c*) By Concrete Sciences those which, like Astronomy, Geology, Biology, &c., deal with distributions and redistributions of matter and motion, molecules, solids, gases, organic phenomena, &c. As to the character of that classification see the criticism on pp. 98-103. Further, M. de la Grasserie has accepted Wundt's distinction of general and special sciences but rejected the distinction of formal and real sciences. The latter, however, if properly drawn, is just as certain and accurate as the former; and it is unfortunate that our author, while recognising how intimately the mathematical sciences are related to the physical sciences, should have failed to recognise that they are related also, although in a lesser measure, to the psychical sciences. Mathematics has undoubtedly a place and function in psycho-physics, human and comparative psychology, economics, ethics (moral statistics), and sociology. How far it will advance it is for the future to decide.

Karl Pearson.

Karl Pearson, the Gershom Professor of Mathematics, has given a classification of the sciences in his well-known work the *Grammar of Science*. The work was published in 1892, and has gone through at least three editions. The classification is only dealt with in the last chapter. The nine chapters which precede it treat of a great variety of subjects bearing on science or sciences, as, *e.g.*, the scope, claims, domain, or method of science; the facts of science; the meanings, progress in formulation, and universality of scientific law; cause and effect, as also probability; space and time; the geometry of motion; matter; laws and life. All those subjects and others are brought by Prof. Pearson before his readers in a most emphatic and vigorous style, and with the utmost faith in himself and in whatever he affirms. Self-criticism, however, is obviously not one of his characteristics, otherwise when writing his *Grammar of Science* he could not have failed to discover that he was really as much of the sort of metaphysician he despised as of the scientist he adored. He begins his chapter on the classification of the sciences with "a summary as to the material of science," and claims for "the heritage of science the whole domain to which the word knowledge can be applied," whereas it is philosophy as *scientia scientiarum* which makes that claim. No single science can reasonably do so, nor even all special sciences combined, as every single science has a definite and limited sphere of its own. Then he reminds his readers again, as he had been doing all through his work, that "knowledge is essentially a description and not an explanation,"—a quite unproved, and probably unprovable, generalisation of Kirchoff's definition, not of all sciences, but merely of Mechanics. Whoever has looked into the *Grammar of Science* must have been struck with the contempt of its author for "the statements regarding force and matter current in all the elementary text-books of science," and his extraordinary faith in such phrases as "science description but not explanation," "conceptual formulæ," "conceptual shorthand," and a host of other questionable phrases. Probably few books will be found less serviceable as an elementary text-book of science than Prof. Pearson's own *Grammar of Science* if used as such, although being in various respects a work of ability it may be very stimulating and useful to those who can separate metaphysics from physics and rhetoric from logic in ways which the author himself has not always succeeded in doing.

As regards the problem of the classification of the sciences, he approaches it with a clear perception of its difficulty, and even with an almost excessive humility. He recognises, to use his own words, "how incapable any individual scientist must nowadays be of truly measuring the importance of each separate branch of science and of seeing its relation to the whole of human knowledge. An adequate classification could only be reached by a group of scientists having a wide appreciation of each other's fields, and a thorough knowledge of their own branches of learning. They must further be endowed with a sympathy and patience enough to work out a scheme of combination."[1] And again he writes: "An individual even with the ability of Bacon or Spencer must fail for want of specialists' knowledge to classify the sciences satisfactorily. A group of scientists might achieve much more, but even their system would only have temporary value as the position of a science relative to other changes

[1] P. 443.

with its development."[1] These are certainly words of soberness and truth. There can be no satisfactory classification of the sciences without careful distribution of them into groups, a comprehensive exhibition of the connections between the groups, and a patient attempt to trace the relationships of the members of each group. The history of the classifications of the sciences is of itself ample proof of that.

Prof. Pearson has taken into consideration only the schemes of Bacon, Comte, and Spencer, imperfect although they be, and expressly tells us that his own scheme, which is derived from these, "pretends to no logical exactness";[2] and that he "is content to call it an enumeration if the logician refuses it the title of classification; for he readily admits that he is not likely to be successful where Bacon, Comte, and Spencer have failed." But surely any scheme of classification should aim at logical exactness; and to aim at surpassing the schemes even of Bacon, Comte, and Spencer need imply nothing presumptuous. The latest scientists have always an advantage over their predecessors. Further, how can a man be reasonably content to call a classification an enumeration, what it is not and cannot be? A mere enumeration of the sciences can only be useless or worse than useless. Prof.

[1] P. 474. [2] P. 452. [3] P. 452.

Pearson's scheme is nowhere merely an enumeration, but everywhere a kind of classification, one mainly composed of three other classifications generally recognised to be far from perfect but also of considerable value.

It is a scheme composed of three sections—viz., A. Abstract Science. *Modes of Discrimination ;* B. Concrete Science. *Inorganic Phenomena ;* and C. Concrete Science. *Organic Phenomena.*

In A the general relations of discrimination dealt with are (*a*) either qualitative and quantitative, as also (*b*) relations peculiar to space and time.

As regards the qualitative relations—*Logic, Orthology* (by which is meant " the study of the right use of language, the clear definition and, if needful, invention of terms), and *Grammar.* As regards the quantitative relations there is a division of *discrete quantity* and another of *change in quantity.* Under the heading 'discrete quantity' *Arithmetic, Algebra, Theory of Measurement, Errors, Probability, Statistics,* &c., and under that of 'change in quantity' *Theory of Functions, Calculus of Rates* or *Functions, Calculus of Sums,* &c., are assigned a place. Connected with the special relations of space are held to be *Descriptive Geometry, Metrical Geometry, Trigonometry, Mensuration,* &c., and with those of time *Theories of Observation* and *Description* (qualitative), as also *Theory of Strains* and *Kinematics* (quantitative). Abstract Science is represented as inclusive of all that is generally considered *Logic* and *Pure Mathematics.*

B. To this second section belong concrete as opposed to abstract science and inorganic as distinct from organic phenomena. The common name for the sciences included in it are physical sciences, and by Pearson they are subdivided into what he calls precise or exact and synoptic or descriptive physical sciences,—the former being held to be those reduced and the latter those not yet reduced to ideal motions. Molar Physics, Molecular Physics, Atomic Physics, and Physics of the Ether are viewed as so many groups of Precise Physical Science. "In Molar Physics," says our author, "we deal with the motion which conceptualises the changes of position in bodies at the surface of the earth, *Mechanics;* with the motion which conceptualises the changes in the planetary system, *Planetary Theory;* and with the motion by which we describe changes in the configuration of a planet and its satellites, *Lunar Theory.*"[1] To Molecular Physics he attaches *Crystallography, Hydromechanics, Aeromechanics, Theory of the Tides,* &c.; to Atomic Physics *Theoretic Chemistry, Spectrum Analysis, Solar* and *Sidereal Physics,* &c.; to Physics of the Ether sundry studies apart from and also in association

[1] P. 461.

with the Molecule, as *e.g.*, *Theory of Radiation, Light, Heat, Electricity, Magnetism*, and *Theories of Dispersion, Absorption, Transmission, Conduction*, &c. The Synoptic Physical Sciences are the *Theory of Inorganic Evolution, Geology, Geography, Meteorology, Mineralogy, Chemistry*, &c.

C. The third and last great field of knowledge according to Pearson is the division of concrete science which deals with organic phenomena. It includes the biological sciences, and he subdivides them into those which deal more especially with space or the localisation of life and those which deal more especially with time or growth. In the first subdivision he places what he calls *Chorology* (geographical distribution of living forms), *Ecology* (habits in relation to situation and climate), and Natural History (in old sense); and in the second *History* as non-recurring and *Biology* as recurring growth. History is further described as comprehending the general evolution of species, connected with which are *Phylogeny, Palæontology, Origin of Species*, &c., and the special evolution of man, connected with which are *Craniology, Anthropology*, &c., as regarding his physique; *Art, Literature, Science*, and *Philosophy* as dependent on his mental faculties; and *States, Laws, Customs, Archæology, Folklore*, &c., as inseparable from his social institutions. There follow *Morphology, Histology*,

Anatomy, *Evolution*, *Theory of Sex*, *Theory of Heredity*, *Physiology*, *Special Psychology of Man*, and *Sociology*, the last branch of psychology, but also one which subdivides into such branches as the *Science of Morals*, the *Science of Politics*, *Political Economy*, and *Jurisprudence*. The whole scheme is brought to a close with *Applied* Mathematics, which link *Abstract Science* to the *Physical Sciences*, and *Bio-Physics*, which connects the *Physical and Biological Sciences*.

Prof. Pearson has candidly acknowledged that freedom from errors cannot be claimed for the foregoing scheme, and certainly the errors of it are numerous. Logic, Orthology, and Grammar are the members of his first group. But of the three only the first is a science. So-called Orthology is merely a portion or function of Logic which almost all books on Logic deal with in some measure, but which it is an abuse of language to designate a science in itself. Further, what is meant by Grammar? and why is it located in the first group of sciences? Is it even Grammar in the ordinary sense of the term? In that case it is nearly equivalent to what Pearson calls Orthology, and there would seem to be no good reason for the invention of the latter term, and still less for counting the same science, if a science at all, twice. Or, Is *Grammar* to be understood in the sense which he attaches to it in the title and throughout the body of his own book?—a book in which he pronounces judgment on the relation of science to theology and metaphysics, as well as on the natures and relationships of causes and effects, matter, motion, life, &c. Epistemology, Logic, and Methodology would, I think, have formed a much more natural group than the one he has given us. As regards most of the other group there is no less room for criticism. The author of them has trusted too much to Bacon, Comte, and Spencer alone; and has apparently not even looked at what, for example, Ampère, Whewell, and Wundt have done in the matter. In the edition in my possession he has not even referred to them. He confidently denies the reality of either theological or metaphysical science. The closing words of his *Grammar of Science* are these: "We have a duty before us, which, if we have faith in the scientific method, is simple and obvious. We must turn a deaf ear to all those who would suggest that we can enter the stronghold of truth by the burrow of superstition, or scale its walls by the ladder of metaphysics. We must accomplish a task more difficult to many minds than daring to know. We must dare to be ignorant. *Ignoramus, laborandum est.*"[1] It is strange that a man of the ability of Prof. Pearson could fancy that by such rash and

[1] P. 474.

random words rational beings would be induced to ignore all theological and philosophical studies as superstition and folly. How daring to be ignorant can be profitable to any mind or any inducement to labour he has not told us and probably cannot. All labour and science presuppose a desire of knowledge. That no one should enter into any burrow of superstition may be readily granted; that all theology is superstition must be proved instead of merely asserted. As to scaling the walls of truth with a ladder of metaphysics a good deal depends on the ladder, and Prof. Pearson may have been unfortunate in the choice of one. I cannot suppose him to be ignorant of the fact that an encyclopædic study, a comprehensive and organic study, of the theological sciences, has had a far longer history than any other group of sciences. The history of it has been continuous through so many centuries, and on the whole so progressive and beneficial, that unprejudiced men are most unlikely to deem all theology a mere "burrow of superstition."

VIII. FROM PAUL JANET TO PRESENT TIME.

From Karl Pearson I must pass to the late Monsieur Janet, a man of very differently constituted mind. During the last half of the nine- **Paul Janet.**

teenth century France had probably no more admirable representative of philosophy than the latter. For almost fifty years a professor of philosophy, he made himself acquainted with all forms, phases, and departments of it; was, to use his own words, always ready "to seek its foundations, authority, limits, and signification, by confronting it with the data and conditions of modern science, as well as with the doctrines of the boldest and most recent metaphysics"; and could most justly say, as he has actually done, *nihil philosophicum a me alienum putavi.* He has written many philosophical works, not one of which is other than valuable, and most of which should long deserve to be studied. His *Causes Finales* (translated into English in 1878) is the best work on the subject. Hardly less important is his *Principes de Métaphysique et de Psychologie*, published in 1897, two years before his death. The first twenty lectures of the first volume of it all bear more or less on the subject of the relations of philosophy and the sciences to one another, and also on the classification of the sciences. To them I must refer.

The first lecture is an admirable discussion of the question, Is philosophy a science? The second is an equally admirable examination of certain modern definitions of philosophy. The third and fourth treat of the criterion of philosophy. And the fifth is an inquiry as to what is or ought to be the respective and appropriate functions of science and belief in philosophy. None of the subjects of those lectures are irrelevant to a study either of the organisation of science or the classification of the sciences, for the simple but almost always overlooked reason that philosophy and science are most closely connected, and that neither can in any form be wholly severed from the other without serious detriment to both. In his sixth lecture Janet gives an account of just five classifications of the sciences—namely, those of Aristotle, Bacon, Ampère, Comte, and Spencer; and the conclusions arrived at are that the classification of Aristotle is antiquated, of Bacon superficial, of Ampère artificial and complicated, of Comte simple and solid but incomplete and mutilated, and of Spencer more comprehensive than that of Comte but also incomplete and likewise burdened with defects justly ascribed to the scheme of Ampère.

In his seventh lecture Janet begins his own attempt at a classification of the sciences, but distinctly refuses to commit himself to presenting a systematic and complete plan such as Ampère and Spencer had endeavoured to provide. He first proceeds to indicate the reasons which had been or may be advanced in favour of a linear series of sciences; and then carefully to show that plausible as those reasons may be there is a fact of such a kind—the fact of consciousness—which when carefully considered makes absolutely incredible to sane reason belief in a merely linear series of sciences. Hence he falls back on the distribution of the sciences into *cosmological and noölogical sciences*, or into *sciences of nature and sciences of humanity.* The sciences of nature or cosmological sciences are subdivided into two classes. As regards the first group, these are the sciences which are concerned with the most general conditions of matter, and specially occupied with measurement, numeration, extension, and motion. Such are arithmetic, geometry, mechanics, and the still more abstract sciences, algebra, and the differential and integral calculus. Astronomy, physics, and chemistry, although less abstract and comparatively concrete, are placed in the same group and treated as abstract and fundamental sciences. Geology and mineralogy, however, are viewed as concrete sciences attached to terrestrial physics. The second group of cosmological sciences are those which treat of life and its phenomena. It also includes abstract and concrete sciences,—those which treat of life in general and those which study living beings. Biology is the science of life in general. As such it subdivides into three great sciences—Biotamy, Biotaxy, and Bionomy. Biotamy corresponds to anatomy, and is the science of the structure of living beings. Biotaxy is the science of the classification of living beings. Bionomy corresponds to physiology, both general and comparative. Botany and zoology are two concrete sciences connected with those that are abstract. The sciences of humanity should follow in due order. They all rest on a fundamental fact, the fact of consciousness, and are divisible into three orders of sciences — (1) Historical sciences; (2) Philological sciences; and (3) Sociological sciences. While distinct from the sciences of nature they are notwithstanding related to them. History, for example, is inseparable from geography, and geography is connected with geology and astronomy. Psychology itself is intimately united with physiology. To psychology as the science of the facts of consciousness lectures eight and nine are devoted. Comte's criticism of the science is shown to have greatly misrepresented it from his desire to get rid of it; and, following his example, some later writers have fallen into errors as to its nature. Janet has done justice to metaphysics by raising in lecture ten such questions as, Is there no other science or class of sciences than those already mentioned? Is there not a science superior to, after, and above any merely particular science? Is there not the science known by the name of Metaphysics from the time of Aristotle to the present day? Is Metaphysics not a legitimate and necessary study in so far as positive philosophy or logic of the sciences? Is it not so far likewise as a synthesis of the universe under the form of philosophy of evolution or any other form? Or as a critique of knowledge? Or as knowledge even of the unknowable so far as in any measure knowable? Or in so far as a final synthesis or as a synthesis of the sciences of nature and of humanity?

The lectures which follow those that have just been noticed are not directly occupied with classification of the sciences, but they have indirect bearings on it of very great importance. The subjects to which I refer are *the relations of theology and philosophy* (lectures 12 and 13), of *philosophy and the sciences* (14 and 15), of *philosophy and history* (16), of *philosophy and geography* (17 and 18), of *philosophy and literature* (19), and of *philosophy and politics* (20). They are all subjects of a kind to be studied and taken into account by those who would aim at a thorough organisation of the sciences,—all of a character indispensable to any one attempting so great a task. By Janet they have been dealt with remarkable clearness and comprehensiveness, and with entire freedom from any kind of prejudice or exaggeration. Although not direct efforts at classification, they must indirectly be most helpful towards a thorough insight into the natures alike of philosophy and science in all their relationships.

Monsieur Edmond Goblot published in 1898 an *Essai sur la classification des sciences*, a work of 296 pages. The spirit of positivism pervades it from beginning to end, although Comte's views and conclusions are often criticised and rejected. M. Goblot endeavours in many instances to be a more thorough and consistent positivist than Comte, and assumes that all philosophical questions and conclusions properly belong entirely to some positive science or other. The assumption is one which facts are not yet found to have verified. No philosophical question properly so called has been shown to belong exclusively to any of the so-called positive sciences. Philosophy always of its very nature transcends more than is attained or attainable by a single exact science. Goblot.

The work of M. Goblot consists of two parts. The first is much shorter than the second, and also of considerably less importance. The title given to it is "The Formal Unity of Science"; and induction and deduction are represented as merely two stages in the development of certain sciences, not as two distinct methods proper to them. All sciences, even the mathematical, — arithmetic, algebra, and geometry, — are maintained to have followed the same method of procedure, one uniform or homogeneous in direction as being alike inductive and deductive; not two distinct methods separate from each other, an inductive which begins with groping, seeking, and finding, and a deductive, synthetic, demonstrative process. The accuracy of that view may not unreasonably be doubted. Possibly such plausibility as it may appear to have may be due to failure on M. Goblot's part to distinguish and separate the two stages of knowledge, ordinary and scientific. Mathematical demonstration belongs exclusively to the latter and higher stage. According to the author of the *Essai*, all true science tends to become abstract and deductive, the experimental as well as the mathematical. That may or may not be so. Considering how far mathematics has during the nineteenth century extended its bounds, what thoughtful and educated man will venture to say where will be its limits at the close of the twentieth? There has probably been nothing more marvellous in the nineteenth century A.D. than the development and expansion of mathematical thought.

The title given by M. Goblot to the second section of his work is "The System of the Sciences," and in that section he subjects to a very close examination the arrangement and classification of the sciences. As was to be expected, he has maintained that of all sciences the mathematical are entitled to have the foremost and dominant place assigned to them. They are, of course, represented as composing the first group of sciences. *Arithmetic*, *algebra*, *geometry*, and *mechanics* are held to be its constituent sciences. The first two, inasmuch as they are occupied not with measurable things like space and motion but with pure quantity, measurement in general, are deemed entitled to be placed before geometry and mechanics. Geometry is placed next in order on the ground that it starts from the idea of space, the conception of extension, what is also directly measurable. Mechanics follows as dependent on the idea of time, and is viewed as including kinematics, the science of movements, and dynamics, the science of forces. According to M. Goblot it is the best example of a science which has become deductive as soon as its elementary notions have been elucidated and its essential definitions formulated. Like all pure science, he holds it to be entirely abstract, and as such altogether independent of the reality of its objects. He denies that the notion of mass is what differentiates kinematics and dynamics, and affirms the real distinction between them to be that the former is concerned only with real motions whereas the latter takes account also of possible motions. The sciences of the mathematical group are said to have no need of resting on experience as they are always in conformity with experience. A fact not in conformity with mathematical laws is an impossibility.

The physical sciences are next brought before us under the head of Cosmology. In their present condition they are, of course, allowed to be experimental and inductive, but they are also affirmed to be destined to become deductive. The following is the list given of them (see ch. iv. pp. 128-156 of M. Goblot's *Essai*) :—

Physics, described as *theoretical and abstract cosmology* and inclusive of various studies—viz., (*a*) the study of the mutual gravitation or attraction of masses (*barology*), (*b*) the study of heat (*thermics*), (*c*) *optics*, (*d*) *acoustics*, and (*e*) *electrology* understood as not only the study of electricity but also of magnetism. The study of molecular actions is also added, but only so far as confined to physics and consistent with physics and chemistry being two quite distinct sciences. *Physics* is defined as *the science of matter*, but *matter* is said to have *no ontological meaning*—*i.e.*, to be not a reality but an abstract conception; and by the indefinite possibility of bodies as space is meant the indefinite possibility of figures. Body is affirmed to be space occupied in opposition to space empty, but the physicist is told that it does not belong to him to say by *what it is occupied;*

that the chemist and mineralogist should be left to determine *that.*

Chemistry, with which mineralogy is intimately connected, is characterised as *special or systematic cosmology.* The new conception here is said to be that of *bodies as actual things.* The body, the elementary body, is the atom. And according to M. Goblot the atom, although indivisible, extended, and impenetrable, has no sensible properties, neither temperature nor colour nor even resistance, neither solidity nor fluidity. He has strangely little to say of it, and virtually nothing of what others have said of it, much and disputed as that has been.

Astronomy and physical geography are characterised as *forms of descriptive, concrete,* and *theoretic cosmology. Cosmogony and geology* are described as *historical, concrete,* and *theoretical cosmology.* The concrete and theoretic are what they are held to have in common. What is pronounced distinctive of them is that astronomy and physical geography are 'descriptive sciences' and that cosmogony and geology are 'historical sciences.'

The last great group of sciences dealt with by M. Goblot is now reached. He has treated it at far greater length than either of the two correspondent groups which preceded it. It is composed of *Biology, Psychology,* and *Sociology,* and designated *Bio-Psycho - Sociologie,* a somewhat clumsy but appropriately comprehensive term, one meant to indicate three important and distinct yet related sub-groups of very important sciences—namely, the biological, psychological, and sociological. The fundamental idea, however, which one would expect to connect all biological, psychological, and sociological science, is the very reverse of clearly brought out. What it is I confess I do not know. Perhaps it may be *the idea of finality*, but if so, there is no definite statement to that effect.

Physiology occupies in our author's scheme almost the same position towards *biology, psychology, and sociology as physics towards cosmology.* As pure and abstract or general physiology it is coextensive with all biology, and is the science of all the laws of life, or more simply the science of life. It is in close connection with *anatomy.* They march side by side. Neither without the other would have attained to the full rank of science. The great stages of progress in physiology have been preceded by discoveries in anatomy, and anatomy without the researches of physiology would be unable to elucidate its own observations. Physiology indeed, as understood by M. Goblot, can only adequately accomplish its work by combining and co-operating with such species of knowledge as histology, embryology, morphology, phylogeny, pathology, and teratology. Zoology he connects with botany, and describes them as systematic or special and applied or concrete biological sciences. Anthropology he includes in zoology. What he calls biological geography he describes as a biology which is descriptive, applied, or concrete, and as a geography which is linguistic, economic, political, &c. He further includes palæontology and history as closely connected in this section of his scheme, and as both occupied with 'the order of facts in time.' Apparently he has overlooked that that is true also of all sociological studies. Hygiene and therapeutics are appended as 'practical sciences,' on the ground that they are serviceable to plants, beasts, and men.

M. Goblot next proceeds to assign to psychology its appropriate position in the scheme of classification of the sciences. He affirms its dependence on physiology and biology, and indicates the relationship between it and them. Further, he endeavours to describe what physical phenomena are and to show their inseparability to some extent from physico-chemical phenomena. As to what psychology itself is, however, he has said disappointingly little, and that little is not of much importance. The comparative psychology both of human races and animal species is entirely overlooked, although it well deserves to be regarded as what it must probably soon become acknowledged to be—viz., one of the greatest and most instructive of sciences. There is a like oversight as to mental pathology.

As a positivist of the Comtist type M. Goblot should bring his classification of the sciences to a close in sociology. Comte did so, and was in that respect self-consistent, but it is not evident that his disciple is so. Comte divided sociology into social statics and social dynamics, the former being the theory of the spontaneous order of human society and the latter the theory of its natural progress,—the one exhibiting the conditions of the social existence of the individual, the family, and the species, and the other the course of human development. What M. Goblot does seems to be something very different. He appends to sociology logic and æsthetics, and thereby implies that logic and æsthetics are of later origin and rank than sociology. True, he speaks of them as the remotest branches of sociology, and thereby implies the latest, but he does not show that they are branches of it at all. The logic of Aristotle, who died in 322 B.C., was at least as great an achievement as the sociology of Comte, although the former preceded the latter by so many centuries. Further, if logic and æsthetics can be so located or characterised as M. Goblot represents them to be, ethics and economics may be equally so, and in that case more may reasonably be said for the priority of them all to sociology than for the priority of sociology to any one of them. It has also to be noted that logic, æsthetics, ethics, and economics form a distinct group of sciences, each of which has a definite aim of its own and a nature akin to but not identical with the others. Thus logic is occupied with the nature, conditions, and processes of reasoning as its subject-matter, and with the attainment of truth and exposure of error as its appropriate ends. Thus beauty is the distinctive object, and the realisation and enjoyment of it the final causes, of æsthetics. So ethics not only undertakes to study men's moral natures, moral relations, and moral histories, but also endeavours to direct and regulate their actions. And similarly, while the specific matter of economics is public wealth, its distinctive ends are the production and distribution of that matter in the most appropriate and socially beneficial manner.

A favourably known Neo-Kantish philosopher, **Stadler.** Prof. A. Stadler of Zürich, published in the *Archiv für Systematische Philosophie* (Bd. ii. 1, N. F., 1896) a contribution to the subject in hand, entitled *Zur Klassification der Wissenschaften.* He had already made an attempt of the kind in 1887. He prefaced his scheme with remarks on the views of De la Grasserie and Wundt, which seem to me of little relevancy and less value. That he should speak of such attempts as "relatively rare" seems to me an astounding statement, which the present volume should amply refute. He starts by defining science as 'the most exact description possible of the totality of the representations given to human consciousness.' That may pass as a harmless statement, but Stadler, following the bad example of some other recent German writers, has talked in such a confused way about what should be meant by the terms "Beschreibung," "Vergleichung," "Mittheilung," "Benennen," "Mittheilen," "Erklären," &c., as tends to the reverse of elucidation. On that subject readers may consult Herr Otto Schneider's review of Stadler's *Klassification.* (See *A. S. Ph.*, iii. Bd. i. 1-19.)

The first and most comprehensive section of sciences in Stadler's scheme of classification is that in which the sciences are divided into those which come under the heading either of *Erscheinungslehre* or of *Ideenslehre*—either into sciences which rest on phenomena or on ideals, on what is or what ought to be. The sciences which have physical and psychical phenomena for their objects and forms are numerous and compose subordinate groups, of which the first and largest is occupied with external and physical phenomena, and entitled *Körperlehre.* The members of this group are classified by Stadler as follows :—

A. I. *Morphology.*—It is said to deal with phenomena and their changes as they are immediately given, and is represented as a generic science of which those others are specific—namely, (*a*) Cosmology, as knowledge of the external phenomena of the universe; (*b*) Astronomy, the objects of which are the celestial bodies; (*c*) *Erdkunde*, such an acquaintance with the earth as includes Meteorology, Geography, and Geology; (*d*) Mineralogy; and (*e*) Biology, conjoined with which are Botany, Zoology, and Physical Anthropology, which all deal with the study of organisms.

II. *Chemistry.*—It is represented by Stadler as dealing with external phenomena that are combinations of elements and as having the following departments belonging to it: (*a*) Analytic Chemistry; (*b*) Synthetic Chemistry; (*c*) Astro-chemistry; (*d*) Geo-chemistry; (*e*) Chemistry of Minerals; and (*f*) Biological Chemistry. As regards Synthetic Chemistry, the syntheses are referred to as either inorganic or organic.

III. *Histology.*—Is occupied with organic phenomena as combinations of vegetable and animal tissues of the smallest and simplest order. It seems questionable that it should be held to precede either Anatomy or Physiology as it does in the scheme, and questionable also that it should be given precedence to Physics.

IV. *Physics.*—Has many sciences assigned to it. It is described as being no less than *synthetic*, *analytic*, *cosmic*, *astrophysical*, *mechanical*, *optical*, *acoustic*, *magnetic*, *electric*, and *thermal*, which means ten sciences in one. But there are added to it four other sciences—Physical Geography, Physics of Minerals, Special Physiology, and Special Psychophysics.

V. *History.*—The objects of history are phenomena and their changes as given at different times and in an orderly succession. Belonging to it are said to be Cosmogony, Astrogeny, History of the Earth, History of Development, Autobiography, General and Special Biography, and the General and Special History of Culture. Certainly not all of these are entitled to be deemed sciences strictly so-called, however interesting they may be as studies.

B. SEELENLEHRE (PSYCHOLOGY).—Is the science of mind and self-consciousness, but also intimately connected with the nature and states of a corporeal organism. Stadler assigns to it the following studies as sciences—viz., (*a*) Subjective Psychology and Autobiography; (*b*) Objective Psychology, Psychophysical Anatomy, General Psychophysics, Special Psychophysics, and General Life-History of the individual consciousness (Special Biography); also (*c*)

Comparative Psychology, Universal History of Culture, and Special History of Culture.

C. SEINSOLLENDE (IDEENLEHRE), knowledge of the obligatory and ideal, subdivides into Teleology, which has to do with happiness, and Ethics, which is conversant with morality.

I. *Teleology* has the following subdivisions: (*a*) Pure Teleology; (*b*) Applied Teleology; (*c*) Eudämonistic Pædagogy; (*d*) Economics; and (*e*) Æsthetics.

II. *Ethics.*—It is subdivided into (*a*) Pure Ethics, which treats of absolute morality; and (*b*) Ethical Pædagogic, which concerns itself with the relationship of appearances to absolute morality.

D. MATHEMATICS.—Stadler regards the mathematical sciences as occupied with *the possible forms of phenomena.* He has contented himself with enumerating merely *three* such sciences—namely, Geometry, Arithmetic, and Kinetics. But are mathematicians likely to be satisfied with so few? Or, are they likely to acquiesce in the three that are mentioned being placed *last* in any classification of sciences? Is it not a fact that they have very generally been accustomed to see their sciences placed in the first rank of most classifications of the sciences?

Stadler's scheme of classification must be credited with containing not a few good points and some admirable suggestions. As a whole, however, it is far from satisfactory, and many earlier schemes are likely to be preferred to it. I have already referred to Schneider's criticism of it.

Trivero. Three years later than the appearance of Stadler's scheme the *Classificazione delle Scienze* of Signor Camillo Trivero was published. It is a work of nearly three hundred pages, and one of the books in the *Collection* of the *Manuali Hoepli*, so termed from the well-known publishing firm in Milan. The book of Signor Trivero has been much influenced by the treatise of M. Goblot that has already been under consideration in this volume. It may suffice to treat it briefly.

Signor Trivero maintains, like M. Goblot, the necessity of classifying the sciences both from an objective and a subjective point of view. All classifications regarded only from either standpoint of observation are held by him to be necessarily very defective. In his opinion, as in his predecessor's, the sciences must be distinguished from one another either by differences of the facts with which they have to deal or by differences of the points of view from which the same facts are contemplated and examined. Differences of method, he holds, are not to be taken into account in any attempts at classifying the sciences. He even denies that there are any such methods. In that respect he has gone farther than M. Goblot, inasmuch as whereas the latter at least admits that there are different methods correspondent to the stages or phases of development in all the sciences, Trivero denies that, properly understood, there are any different methods. There is "only one," he affirms, "only one that is good and scientific, the method which proceeds from the known to the unknown; and it is of little consequence whether that method ascends and is called induction, or descends and is said to be deduction, or proceeds horizontally and is termed analogy." These so-called methods he denies to be distinct methods.

In the opinion of Trivero a system of the sciences should be presented under the form, as M. Goblot has said, "d'un tableau à double entrée, avec 'divisions horizontales' et 'divisions verticales.'" In that respect Trivero and Goblot are agreed, but neither of them has worked out a scheme of the kind to either order or completeness, and Trivero least so, as he has presented no justification whatever of the "horizontal divisions." Holding all knowledge to be capable of being studied from three points of view, he should have shown what the results would be, but that he cannot be said to have successfully done. The three points of view are affirmed to be the historical, scientific, and philosophic, and intimately connected with them are held to be 'vertical divisions' of the sciences. It does not appear, however, from Trivero's scheme that almost any sciences of any kind are to be seen from his 'points of view' or arranged in his 'divisions.' The so-called first point of view is 'history,' and in history 'geography' is included, but not more than 'history' is included in 'geography.' Further, history began its course not as 'science' but as 'art,' as 'literature,' and still is often that and no more. Gradually indeed it passed into a political stage, and even exercised much political and social influence. Later it ceased to be satisfied with merely describing or recording historical actions and events, and sought for a full comprehension and explanation of them. It thus passed into the stage of theoretical and explanatory science, but with only a very slight addition to the number of sciences. Beyond the scientific stage there is admitted to be a philosophical stage, but there is no mention of philosophical sciences, and could not be expected to be, as for Trivero all philosophy is merely metaphysics, and all metaphysics is merely a search for the absolute. Thus far the sciences exhibited must be admitted to have been exceedingly few.

But a hierarchy of the sciences resting on the natural objects and natural sequences of those sciences has still to be recognised, and according to Trivero there are seven of them. The *first* is said to be *Astronomy,* and to have for its object *the sidereal world;* the *second* to be *Geology,* with *the earth* for its object; the *third Mineralogy,* which treats of *the mineral kingdom;* the *fourth Botany,* which is occupied with *the vegetable world;* the *fifth Zoology,* to which *the animal kingdom* belongs; the *sixth Psychology, in so far as man is more than a mere animal;* and *the seventh Sociology, the science of man's actions and productions.* That may well seem to some a very clear and simple distribution of the sciences, or at least of a 'vertical section' of them, but it is certainly also a very inadequate scheme of classification of the sciences as a comprehensive system in which are many members at once distinct and related. Could there have been a science of astronomy worthy of the name of science had there not been prior to it logic, mathematics, and so far physics? If geology be pronounced a science why should geography not? Can mineralogy be a science if chemistry be ignored? Is the definition given to sociology one of which any sociologist would approve? Certainly not. It would be nearer to a definition of anthropology. All sociologists are aware that to define sociology is a very difficult affair.

In 1898 a J. G. Meyer published at Strassburg a book or essay bearing the title *Das natürliche System der Wissenschaften.* I am, however, quite ignorant of its character or contents, having been unable to obtain a copy, notice, or review of it, or even to find out the name of its publisher.

From 1866 to his death in 1901 Monsieur J. P. Durand (de Gros) devoted himself to the study of classification with more zeal, perhaps, than any one in France or elsewhere, while deploring that even naturalists and logicians had contributed exceedingly little towards the development of what seemed to him might be, and ought to be, made a complete and well-established science of universal classification or orderly arrangement in every direction,—the science to which he has given the appropriate title of Taxinomy. The most important of his works, perhaps, is the one entitled *Aperçus de Taxinomie Générale,* published at Paris in 1899 (by F. Alcan, *éditeur*, pp. 265). Too modestly he described himself as merely a pioneer in a region where he had really laboured for almost a lifetime, and seems to have found in it much which alike his predecessors and contemporaries had overlooked. A more earnest and independent treatment of it there could scarcely be. A vainer man who had done as much would have not unlikely claimed to be the author of a *Scienza Nuova.* Yet few readers, and especially readers outside of France, would, I fear, be likely to do justice to such works as those of M. Durand, suggestive and instructive although they be. It seems desirable, therefore, to state that of the *Aperçus* there are two good yet brief notices which will be no great burdens on their readers. One is that of Monsieur F. Paulhan in the *Revue Philosophique* for April 1899 (pp. 419 - 424). The other is that of Prof. Bosanquet in *Mind* for October 1899 (pp. 531-535).

Both reviewers have naturally dwelt chiefly on the main subjects of the works reviewed,—those which Durand himself called the Four Taxinomic Orders or Problems. The First Order is described as that of *Generality or Resemblance.* The classifications of botany and zoology are applications of it, specially included in it, and familiar to us in the relationship between genera and species. Induction, generalisation, and specification are processes implied in it. The entire order is based on the relationship of genus to species and of species to genus. Not so the Second Order,—the order of *Composition or Collectivity.* It is founded on the relation of whole to part and part to whole, and has for its objects concrete objects, not abstract conceptions like those in the first order. The Third Order is the order of *Hierarchy or Relationship of Rank.* It is maintained to rest on relationships of subordination, as, *e.g.*, of superiority, equality, and inferiority. The Fourth Order is that of *Genealogy and Evolution*, and is represented as dependent on affinities of kinship under the three species of *ascent, collaterality, and descent.* Taxinomy was Durand's great contribution to classification, and it was with classification as a whole that he felt himself bound to see it as far as possible fully developed. A classification of the sciences was accordingly not overlooked by him. But he cannot be said to have given it any special attention. It would appear as if it were regarded by him as a comparatively small affair, the settlement of which could only be attained through a rational evolution of the science of taxinomy itself. Study, he seems to have thought, the variations of all the objects and methods of the objects and relations of the sciences, and you will necessarily learn to classify the sciences aright, although so many have failed to do so. We can understand, therefore, how, although he dealt to some extent in the last chapter of the *Aperçus* with the classification of the sciences, it was to a very small extent, and led to no result of consequence.

There is no apparent likelihood of there being fewer attempts at classifications of the sciences in the present than in the past century. No one classification of the kind has yet been generally adopted. The number of sciences to be classified seem to be on the increase, and some of them are difficult both to define and locate. New sciences are generally found to speedily introduce others. An active philosophy is sure to agitate questions which call for settlement from sciences that had previously been dormant or ignored. The great increase of interest shown by scientists of late in classification itself is of itself evidence that classifications of the sciences will not decrease but increase in number. Both taxinomy and morphology are obviously working in that direction under the belief of those who cultivate them that each science is to be carefully assigned to its appropriate position in an appropriate class. It does not follow that a correct and adequate classification of the sciences will be either easily or speedily found. It will certainly not be found in any single linear series. It is much more complicated than that, and seems to be always becoming more complicated. The older sciences are at least as fruitful as they ever were, and the newer sciences are now seldom regarded with suspicion, but, on the contrary, rapidly adopted and warmly welcomed.

Consider for an instant the positions occupied by those three recent and very interesting and flourishing sciences called Anthropology, Ethnology, and Sociology. There are, perhaps, none which have come more rapidly to the front or attracted more attention. Nor are there almost any sciences which have taken possession of vaster regions or more numerous provinces. But they so interlap one another at all points, and so overspread ground claimed by all of them with almost or altogether equal rights, that it is difficult to say what are their external limits or internal contents. So far as they have hitherto been dealt with, any one of them would seem to be largely occupied in attempting to supplant the other two, while professing to be entirely co-operating with them.

Anthropology is a real and very important science, the success of which has been great and well-deserved owing to the labours of its many zealous students. In the United States of America alone there are about forty Universities, and in the majority of them several teachers of the science, anthropological museums, and various means of practical anthropological study. Great Britain and Ireland are not so advanced owing to their want of encouragement and support, but individuals have amply shown how much they could do, and how much more with ampler means might be done. There is happily one admirable institution in the kingdom devoted to the study of anthropological science, and which is well known to have an admirable organ in the *Journal of the Anthropological Institute of Great Britain and Ireland*,—a large annual volume now in the twenty-fourth year of its existence.

Its eminent president, Dr A. C. Haddon, delivered on January 27, 1903, a very interesting address on *Anthropology*, Its Position and Needs. But the very opening sentences of his address are these:—"A peculiarity of the study of Anthropology is its lack of demarcations: sooner or later the student of Anthropology finds himself wandering into fields that are occupied by other sciences. The practical difficulty of drawing a dividing-line between the legitimate scope of Anthropology and that of other studies is so great that we are often told there is no science of Anthropology. This lack of definiteness adds a charm to the subject and is fertile in the production of new ideas, for it is at the fringe of a science that originality has its greatest scope. It is, however, only by a synthesis of the various studies which are grouped together under the term Anthropology, that one can hope to gain a clear conception of what man is, and what he has done."[1] And he adds: "It may be logically consistent to distribute portions of Anthropology among other sciences, but

[1] Vol. xxxiii. p. 11.

the result would be that the subject would suffer, and unless a society like our Anthropological Institute busied itself with the study as a whole, it would be developed very unequally. Indeed, to be quite candid, at the present there is very little direction in the evolution of Anthropology, or in the study of its branches."[1] He has further drawn out, with the fully acknowledged co-operation of Professor Patrick Geddes, a very remarkable scheme of classification of sciences, or at least of studies, all held to belong to, and even to be portions of, Anthropology. The scheme is represented as having three planes. The lowermost plane may be designated anthropological and even biological. Adherent to it are held to be the following sciences, and they are arranged in two parallel series thus :—

I.	2. Palæontology.	Taxonomy.	Œcology.	Rational Phylogeny.
	1. Embryology.	Anatomy.	Physiology.	Rational Ontogeny.

The two series are included in the first and lowermost plane. Series number one is the lowest of all in the scheme, and on the whole lies beyond the legitimate bounds of both Anthropology and Sociology, Taxonomy and Anatomy. Neither Taxonomy nor Anatomy belongs exclusively or distinctively to Man. The same is true of Physiology.

[1] Vol. iii. p. 11.

Œcology is an ambiguous term, being employed by some, Hæckel for example, to denote the science of Economics as applied to plants and animals, and by others to *nature-folk* as distinguished from *culture-folk*. The term Ontogeny is employed in biology and psychology for individual development, as contrasted with the term Phylogeny, which is used to denote the process of the descent and development of species, and to explain the ancestry and genetic relations of organisms.

The second plane with its two parallel series are manifestly more entitled to be regarded as of an anthropographical or anthropological stage than the first. It is, however, arranged just in the same way. It is the intermediate stage or plane, and its two parallel series are the following:—

II.	4. Palæontology of Man.	Racial Classification of Man.	Anthropographical Œcology.	Rational Phylogeny.
	3. Comparative Human Embryology.	Comparative Human Anatomy.	Comparative Human Physiology.	Rational Ontogeny.

The man who appears at that stage is truly man, although far from fully man. He is not a mere animal but a social and rational being which has occupations and institutions of the kind distinctly recognisable as human.

On the third plane man is seen to have risen,

aided by the stepping-stones of the stages I. and II., or what is equivalent, those of the series 1, 2, 3, and 4. It is the plane or stage on which, as President Haddon says, "the limitations of the classification in the animal plane are largely transcended," "all the enterprises of social man studied," and where "Psychology takes us into the inner sanctuary of man, and while it, too, has its roots in his animal nature, it flowers, so to speak, in a realm of its own. In the third stage, the uppermost stage, Ethnology and Sociology are identified without proof given. They should be treated as distinct. The uppermost plane is the last, and composed of the two following series of sciences, or supposed sciences, thus :—

III.	6. Archæology.	Social Taxonomy.	Economics and Politics.	Philosophy of History.
	5. Evolution of Institutions.	Analysis of Institutions and Technology.	Functioning of Occupations and of Institutions. Linguistics.	Criticism of Institutions.

President Haddon and Professor Geddes have presented as a whole the planes, sections, and series of their scheme of classification. That, it seems to me, may be found rather too difficult for general comprehension. Therefore I have also distributed it into parts and sections, without the

As Ethnography corresponds, or at least should correspond, to Anthropography, so should Ethnology to Anthropology. Ethnography is merely the descriptive study of all ascertainable groups of peoples. Ethnology is in a stricter sense a science, although one intimately connected with and greatly aided by Ethnography. The latter is occupied with the observation of human groups and organisations, of hordes, clans, races, peoples, and nations, or, in other words, with the status, occupations, and institutions of mankind, whereas the former aims at carrying out the fullest possible investigation and explanation of all that Ethnography may have discovered and described. Keane's *Ethnology* is an admirable exposition of the science so called. It deals in a singularly lucid style alike with *the fundamental ethnical problems* and the *primary ethnical groups.* The work issued from the Cambridge University Press in 1896. As regards the accumulation of ethnographical and ethnological facts and theories, perhaps the *Zeitschrift für Ethnologie: Organ der Berliner Gesellschaft für Anthropologie, Ethnologie, und Urgeschichte*, founded in 1869, has not been surpassed, owing doubtless to having started with the support of such indefatigable workers as A. Bastian, R. Hartmann, and R. Virchow.

Sociology is an advance on Ethnology, as Ethnology is on Anthropology. It has often been referred to in this work in connection with the views of it given by Comte, Spencer, J. S. Masaryk, De Roberty, and many others. Very opposite views of it are still given by equally able men. For instance, Prof. Giddings, a most distinguished American thinker and economist, published in 1897 his *Principles of Sociology*, a work in which the nature of sociology as a science, of its place among the sciences, of its appropriate method, its territory, and distribution of parts, &c., were most skilfully exhibited. In the same year, however, a very subtle and elaborate attempt was made by Prof. Hyslop of Columbia University, in a *Supplementary Number* of the *American Journal of Sociology*, to refute the views of his predecessor. There he dealt with Prof. Giddings' classification in detail, and examined and criticised a number of possible systems regarding the relations between Sociology and all its cognate and auxiliary sciences, or sources of knowledge.

In America, and all the chief countries of Europe, Sociology has now attracted to itself a wide, vivid, and growingly increasing interest. Perhaps its importance has been most adequately realised in the United States, where it has been taught in almost all their Universities, and in a generally independent and practical way. Britain must be admitted to have lagged behind, but has now seemingly awakened up to its duty and interests in the matter. The newly formed Sociological Society starts on right lines, and promises to be worthy of what it should be. It is to be hoped that it may have, as so many other countries already have, an appropriate literary organ for such a science as Sociology is. Of such an organ the *Année Sociologique*, founded in 1896, and since then till now directed by M. Durkheim and an able body of *collaborateurs*, seems to be an excellent model. The distribution of the matter in it appears to be about as appropriate as possible. Little that is relevant to what Sociology is seems to escape the sociological net, or to fail to find in it something that may be of use. The classification in the *Année* is from its first year (1896-1897) to its present year (1903-1904) scarcely at all altered,—a fact which shows that the scheme had been maturely conceived from the first. An "Analysis of the Sociological Literature (in Books and in Periodicals) summarised in the *Année Sociologique* for 1902" will be found at the close of a very valuable paper by Mr Victor V. Branford, "On the Origin and Use of the Word Sociology, and on the Relation of Sociological to other Studies and to Practical Problems." The great variety of classifications of the contents of Sociology to be found in books and pamphlets at the present time should not be regarded as in any way disproving or discrediting the validity and worth of Sociology. It shows merely that Sociology as the general social science is an extremely comprehensive science when compared or contrasted with the special social sciences which are occupied with the composition, elements, and internal organisation of social groups within comparatively limited spheres. There are many 'approaches,' as Prof. Geddes says, to Sociology. There are likewise many sections, and also sub-sections, each of which has its own special characteristics, and depends on distinctive phenomena (statistical, physical, organic, psychical, anthropological, ethnological, or theological), yet which none the less belong to Sociology itself.[1]

I must now hasten to a close. My history of the classifications of the sciences may be said to be ended, and a few concluding words are all that seem called for.

I have not meant the book to be more than what its title means, and I have brought the history contained in it down to the present time. That that history is needed, no one, I think, for whom it has been intended, can fail to acknowledge. It is meant only for a certain class of persons, and

[1] The most comprehensive study of the nature, methods, and aims of Sociology is the *Sistema di Sociologia*, 1901 (pp. 664), of Errico De Marinis, Professor in the University of Naples and Parliamentary Deputy.

whether the class be a large or a small one I do not profess to know. It is a history brought down to a given date, or, practically speaking, the present day. I do not pretend to have succeeded in collecting and dealing with all classifications of the sciences, but I hope to have come nearer than any one else to success in that respect. I have little doubt that of those who take up the book into their hands there will be a considerable proportion who deem its chief fault to be that so very many schemes of classification are presented in it. That criticism or objection will not touch me at all. A selection of comparatively interesting classifications is not needed, and it can be of very little worth to any one who wishes to have an historical view of the process of classification of the sciences.

While I am writing these lines there is being held at St Louis, U.S.A., a Universal Exposition,—an International Congress of Arts and Sciences,—the express object of which is "to discuss and set forth the unification and mutual relations of the sciences, and to thus overcome the lack of relation and harmony in the scattered specialistic sciences of our day." There has never, so far as I am aware, been known in the history of the world any such event in the history of classifications of the sciences, and if that event be a success the latter history,—

the history of classifications of the sciences,—far from being ended or drawing near to a close, must receive an altogether exceptionally powerful progressive impetus.

Considering the character of the arrangements and the qualifications of those to whom they are intrusted, there is every likelihood that the event will be a great success.

THE END.

PRINTED BY WILLIAM BLACKWOOD AND SONS.

HISTORY, PHILOSOPHY AND SOCIOLOGY OF SCIENCE

Classics, Staples and Precursors

An Arno Press Collection

Aliotta, [Antonio]. **The Idealistic Reaction Against Science.** 1914

Arago, [Dominique François Jean]. **Historical Eloge of James Watt.** 1839

Bavink, Bernhard. **The Natural Sciences.** 1932

Benjamin, Park. **A History of Electricity.** 1898

Bennett, Jesse Lee. **The Diffusion of Science.** 1942

[Bronfenbrenner], Ornstein, Martha. **The Role of Scientific Societies in the Seventeenth Century.** 1928

Bush, Vannevar. **Endless Horizons.** 1946

Campanella, Thomas. **The Defense of Galileo.** 1937

Carmichael, R. D. **The Logic of Discovery.** 1930

Caullery, Maurice. **French Science and its Principal Discoveries Since the Seventeenth Century.** [1934]

Caullery, Maurice. **Universities and Scientific Life in the United States.** 1922

Debates on the Decline of Science. 1975

de Beer, G. R. **Sir Hans Sloane and the British Museum.** 1953

Dissertations on the Progress of Knowledge. [1824]. 2 vols. in one

Euler, [Leonard]. **Letters of Euler.** 1833. 2 vols. in one

Flint, Robert. **Philosophy as Scientia Scientiarum and a History of Classifications of the Sciences.** 1904

Forke, Alfred. **The World-Conception of the Chinese.** 1925

Frank, Philipp. **Modern Science and its Philosophy.** 1949

The Freedom of Science. 1975

George, William H. **The Scientist in Action.** 1936

Goodfield, G. J. **The Growth of Scientific Physiology.** 1960

Graves, Robert Perceval. **Life of Sir William Rowan Hamilton.** 3 vols. 1882

Haldane, J. B. S. **Science and Everyday Life.** 1940

Hall, Daniel, et al. **The Frustration of Science.** 1935

Halley, Edmond. **Correspondence and Papers of Edmond Halley.** 1932

Jones, Bence. **The Royal Institution.** 1871

Kaplan, Norman. **Science and Society.** 1965

Levy, H. **The Universe of Science.** 1933

Marchant, James. **Alfred Russel Wallace.** 1916

McKie, Douglas and Niels H. de V. Heathcote. **The Discovery of Specific and Latent Heats.** 1935

Montagu, M. F. Ashley. **Studies and Essays in the History of Science and Learning.** [1944]

Morgan, John. **A Discourse Upon the Institution of Medical Schools in America.** 1765

Mottelay, Paul Fleury. **Bibliographical History of Electricity and Magnetism Chronologically Arranged.** 1922

Muir, M. M. Pattison. **A History of Chemical Theories and Laws.** 1907

National Council of American-Soviet Friendship. **Science in Soviet Russia: Papers Presented at Congress of American-Soviet Friendship.** 1944

Needham, Joseph. **A History of Embryology.** 1959

Needham, Joseph and Walter Pagel. **Background to Modern Science.** 1940

Osborn, Henry Fairfield. **From the Greeks to Darwin.** 1929

Partington, J[ames] R[iddick]. **Origins and Development of Applied Chemistry.** 1935

Polanyi, M[ichael]. **The Contempt of Freedom.** 1940

Priestley, Joseph. **Disquisitions Relating to Matter and Spirit.** 1777

Ray, John. **The Correspondence of John Ray.** 1848

Richet, Charles. **The Natural History of a Savant.** 1927

Schuster, Arthur. **The Progress of Physics During 33 Years (1875-1908).** 1911

Science, Internationalism and War. 1975

Selye, Hans. **From Dream to Discovery: On Being a Scientist.** 1964

Singer, Charles. **Studies in the History and Method of Science.** 1917/1921. 2 vols. in one

Smith, Edward. **The Life of Sir Joseph Banks.** 1911

Snow, A. J. **Matter and Gravity in Newton's Physical Philosophy.** 1926

Somerville, Mary. **On the Connexion of the Physical Sciences.** 1846

Thomson, J. J. **Recollections and Reflections.** 1936

Thomson, Thomas. **The History of Chemistry.** 1830/31

Underwood, E. Ashworth. **Science, Medicine and History.** 2 vols. 1953

Visher, Stephen Sargent. **Scientists Starred 1903-1943 in American Men of Science.** 1947

Von Humboldt, Alexander. **Views of Nature: Or Contemplations on the Sublime Phenomena of Creation.** 1850

Von Meyer, Ernst. **A History of Chemistry from Earliest Times to the Present Day.** 1891

Walker, Helen M. **Studies in the History of Statistical Method.** 1929

Watson, David Lindsay. **Scientists Are Human.** 1938

Weld, Charles Richard. **A History of the Royal Society.** 1848. 2 vols. in one

Wilson, George. **The Life of the Honorable Henry Cavendish.** 1851